First published in the UK in 2018 by Alex Nail
Photographs and text copyright © 2018 by Alex Nail
Foreword copyright © 2018 by Chris Townsend

All rights reserved. No part of this publication may be reproduced, distributed, or transmitted in any form or by any means without the prior written permission of the publisher, except in the case of reviews and other uses permitted by copyright law. For permission requests contact Alex Nail.

Designed by Studio Renton
Printed in the UK by Johnsons of Nantwich Ltd. Printers since 1827

A catalogue record for this book is available from the British Library.
ISBN 978-1-9164424-1-2

Alex Nail is an internationally recognised landscape photographer based in the Southwest of England. He is a qualified mountain leader and runs photography workshops catering to all abilities. All the images within this book are available as fine-art prints.

Alex Nail - alex@alexnail.com - www.alexnail.com

8643337
Printed on Carbon Captured paper

NORTHWEST

Alex Nail

CONTENTS

Foreword   6

Introduction   8

The Coulin Hills   12

Torridon   38

The Great Wilderness   72

Assynt and Coigach   108

Acknowledgements   146

FOREWORD

by Chris Townsend

The Northwest Highlands of Scotland is a land of glorious splendour, a land unlike any other. Strange mountains, steep and rocky, rise out of a mass of sparkling lochans, each one separate and distinct. Liathach, Slioch, An Teallach, Suilven, Stac Pollaidh, Quinag - their names are redolent of far distant times. A mix of Norse and Gaelic, the names conjure up images of Vikings sailing down the coast followed by the Gaels from Ireland settling on the land between the hills, hunting the deer, fishing the lochs. Encompassing Torridon, Fisherfield, Coigach and Assynt this is a region of unsurpassed wildness and dramatic beauty.

In the southern part of the region the mountains are closely packed together, giants standing side by side above a watery landscape below. As you progress north they spread out and eventually reduce in size, isolated peaks amidst vast areas of lochs and low moorland.

I've spent many days and weeks climbing the hills and exploring the vast and complex 'cnoc and lochan' - knoll and pool - landscape that surrounds them. From the summits in the far north you look out on a glittering world of rock and water where the pools and streams and lakes look more dominant than the land until in the distance the latter rears up again to another wedge of high rock. In Torridon and Fisherfield the mountains are more immediate and you are taken by the sheer grandeur of the landscape.

This land feels ancient, as though the bones of the earth are showing through. The history of the area was revealed when two Victorian geologists, Benjamin Peach and John Horne, worked out how earth movements had forced older rocks over newer ones along a line now known as the Moine Thrust. Their contribution to geological understanding is recognised with a monument which stands on the shores of Loch Assynt.

The lifetime of these mountains extends far beyond the brief span of humanity, their origins lying aeons ago when the deep red Torridonian Sandstone of which they are mostly formed was laid down in an ancient sea. The rocks on which they stand are older still. At three billion years old Lewisian Gneiss is one of the oldest rocks in Europe. The basics of the geology can be studied at a nature trail at Knockan Crag. Here at one point your hand can span 500 million years of geological time. The importance of this landscape led to the creation of the Northwest Geopark, the first in the UK.

For the mountain lover this is a landscape unsurpassed but you must first make the effort - this is not easy country. Paths twist and turn, boggy and rocky in equal measure. Ascents are steep and rough, sometimes after long approaches. This is an unforgiving land exposed to Atlantic gales and with little shelter or respite, but the rewards are immeasurable.

Given the harshness of the landscape and the difficulties it poses, to produce a photographic record of the area, one that goes deep into its heart and far from the beaten path, is a tremendously demanding task. Alex took on this challenge and has produced a book of glorious images that show this land in all its moods. Year round he has carried his equipment up to the summits, camped overnight, braved wild weather, and returned over and over again. As well as superb photographic skills good mountaineering ability is required - Alex has both. The result of his efforts is a mouth-watering collection of photographs that is both inspiring and invigorating.

INTRODUCTION

APPROACH | Setting out to make this book was a daunting task, but an enjoyable one. Ever since my first visit to Assynt, I've been drawn to the remarkable landscapes of the Northwest Highlands. Each trip since has fostered a greater appreciation of the unique charms of the four areas that form the chapters of this book. These photographs are by no means comprehensive but I have tried to be ambitious in order to create as thorough a photographic account as possible.

Although I was a photographer first and backpacker second, I now consider myself an equal part of both tribes. Backpacking, particularly when summit camping, has come to be a great love of mine. It allows total immersion in the outdoors with worries left at home, forgotten. The landscape and weather, regularly ignored in modern life, become my world for a short time. I begin to reconnect with nature.

In the course of this project I have hiked up dozens of mountains, sometimes on multiple occasions, throughout the seasons and in all weather conditions. Often this has involved walking many miles with a heavy rucksack and camping out on summits. On the surface this approach might seem an unnecessary hardship, but it was in many ways an essential part of the process. The only bearable way to be on top of a peak for a 4.30am sunrise is to sleep on the mountain. The majority of images in this book were made in this way. In fact, outside of the winter months, it features just eight photographs captured following a night in accommodation.

This book displays none of my failures, none of the wasted trips, none of the days spent at the Torridon Inn waiting for the weather to clear. Seen individually, certain scenes in the book might be considered rare or even lucky. Together they present an almost impossible image of the Highlands, of flaming sunsets, dramatic storms and fresh blankets of snow. This is an inevitable outcome of weeks spent immersed in the mountains. The light changes. Opportunities arise. The numerous failed attempts and wasted trips seem to make sense now - they have become a footnote.

Whilst at times backpacking has taken its toll and I've wished I was at home in more comfortable surroundings, it has far more regularly been life affirming, dusted with unforgettable experiences. Even the worst moments, several of which are recounted in this book, have brought a level of satisfaction and self-worth with them - enjoyment comes afterwards.

The physicality involved should also be put into perspective. Scotland's mountains are small, the peaks in the Northwest rarely exceed one thousand meters and they would barely register on a world stage - this is not the Himalayas. Climbing these mountains and even camping on them is no great feat of human achievement, but it is an adventure. It is strange to think that relatively few people will discover this enjoyment that has become such a large part of my life.

This book is in reality a year late. Successive failed winters left me with a collection of images almost entirely devoid of snow and I was forced to delay. Yet in the course of this project good fortune has followed me more than bad luck and the winter of 2017/18 was magical. I was in Torridon in December and January for fresh snowfalls, it was a winter wonderland. The spring and summer that followed were fantastic too, providing a wealth of dramatic light. Ultimately some of my favourite photographs were captured at the last opportunity.

THE PHOTOGRAPHS | I am part of a wave of photographers who learnt the craft purely digitally having never set foot in a darkroom. Digital has many advantages, but in the context of my own work the ability to work fast and gain instant feedback has proven critical in capturing transient moments of light, particularly in difficult weather on top of a mountain. Gone are the days when photographers have to cart around large wooden cameras with the painstaking setup time they entail.

Digital photography brings with it almost unlimited freedom, particularly when it comes to post production. This is both a blessing and a curse. In the public lexicon 'Photoshop' has become synonymous with fakery. Social media's appetite for attention has encouraged photographers to create increasingly fantastical worlds that no longer represent the landscape. Often this is justified by simply pointing out that photography is art, but it is my belief that photography can be more than art, reality is powerful. Photography can transport the viewer.

The images in this book are digitally edited, but true both to the scene and the experience. Nothing is added or taken away and although alterations have been made to colour and contrast they remain honest depictions. Photography's direct connection to reality is unique amongst the visual arts and it is something I will always value and protect.

On the creative side my images are generally literal rather than interpretative. Whilst many photographers look to abstract the landscape to create mystery, I prefer a more direct approach. My love of photography comes out of a passion for the subject, rather than the creative possibilities of the medium. It's my goal to make myself and the photographic process invisible to the viewer so that the landscape can be enjoyed unimpeded. These images then have the potential to be experienced with spirit of discovery, revealing landscapes and perspectives only known to keen hikers.

I greatly prefer simplicity over complexity. I admire photographers who manage to produce beautiful images from impossibly complicated scenes, but there is no doubt that I favour the clarity that comes from an image reduced to its core components. Although I look for foreground elements to provide depth in my images they must be interesting or beautiful in their own right, they must justify their place in the image. As a result many of the photographs in this book are 'just' mountain views shown in their simplest form. That is the context in which I hope they can be seen.

Above all I hope these images can convey my love of the mountains of the Northwest Highlands. If they can in any way express the sense of wonder that I have experienced in the course of this project then it will have been worthwhile.

THE NORTHWEST HIGHLANDS | Geography has never been my strong suit. Whilst I have an interest in the geological and glacial processes that formed the Highlands, the mountains appeal to me mostly for their beauty and adventure. I have relatively little concern for whether a peak lies in a particular historical parish and, thanks to the Scottish right to roam, I have minimal interest in who owns the land. So it is that my understanding of these mountains primarily comes through my experience of hiking up them and on through the glens that separate them.

The Northwest Highlands are loosely defined as the wide ranging mountains north of the Great Glen. However the much smaller region covered by this book is better described geologically. From Glen Carron up the west coast to Cape Wrath lies a band of Lewisian Gneiss topped with Torridonian Sandstone that is unique on the mainland. It gives rise to mountains that are both distinctive and beautiful.

I have chosen to break up the book into four chapters in accordance with how I have come to know the mountains. I had originally intended to have an additional chapter covering the area up to Cape Wrath incorporating the peaks of Ben Stack, Arkle and Foinaven. Regrettably my early forays were largely unsuccessful and the rolling hills felt a little out of place in the context of the more pronounced peaks to the south.

So four areas remain: The Coulin Hills, Torridon, The Great Wilderness, and Assynt and Coigach, each of which I have attempted to photograph extensively.

From a photographic standpoint this mountain landscape is a wonder. Unlike most of the Scottish Highlands the mountains here stand apart from one another. There are no grand scale ridges connecting them and as a result both late and early light, prized by landscape photographers, strikes them throughout the year. The sandstone geology and glacial past give rise to vast ice-carved corries. On the summits themselves the sandstone produces wonderful rounded forms and precarious towers. When weathered by time the rock decays to sandy soil upon which thick grass can grow - far preferable to the broken scree found on igneous mountains to the south. In the valleys lie winding rivers, countless lochans, and remote freshwater beaches. Dotted around the shores of the lochs and rivers are stands of native Scots Pine and birch and the signs of glaciation are everywhere, echoing an ancient past. The mountains' position near to the coast creates dramatic changes in weather and lighting as fronts sweep in off the Atlantic.

There is endless potential. Photographing these mountains will be a lifetime's work.

Alex Nail, September 2018

THE COULIN HILLS

From Loch Carron to Loch Torridon

Glen Carron forms the southern boundary of the Torridonian Sandstone and the beginning of a rugged mountain landscape. The area is broken down into three main mountain groups, the Coulin Forest, The Ben Damph Forest and the Applecross Peninsula, but many of the mountains belong to what is generally referred to as the Torridon Hills. These peaks are a little less dramatic than their northern neighbours and as a result see far fewer hikers. There are three Munros in the area, Maol Chean-Dearg, Sgorr Ruadh and Beinn Liath Mhòr. However, the best known peak is Ben Damph which offers commanding views of the region. Out to the west on the Applecross Peninsula lies the extended ridge of Beinn Bhàn with its impressive east facing corries.

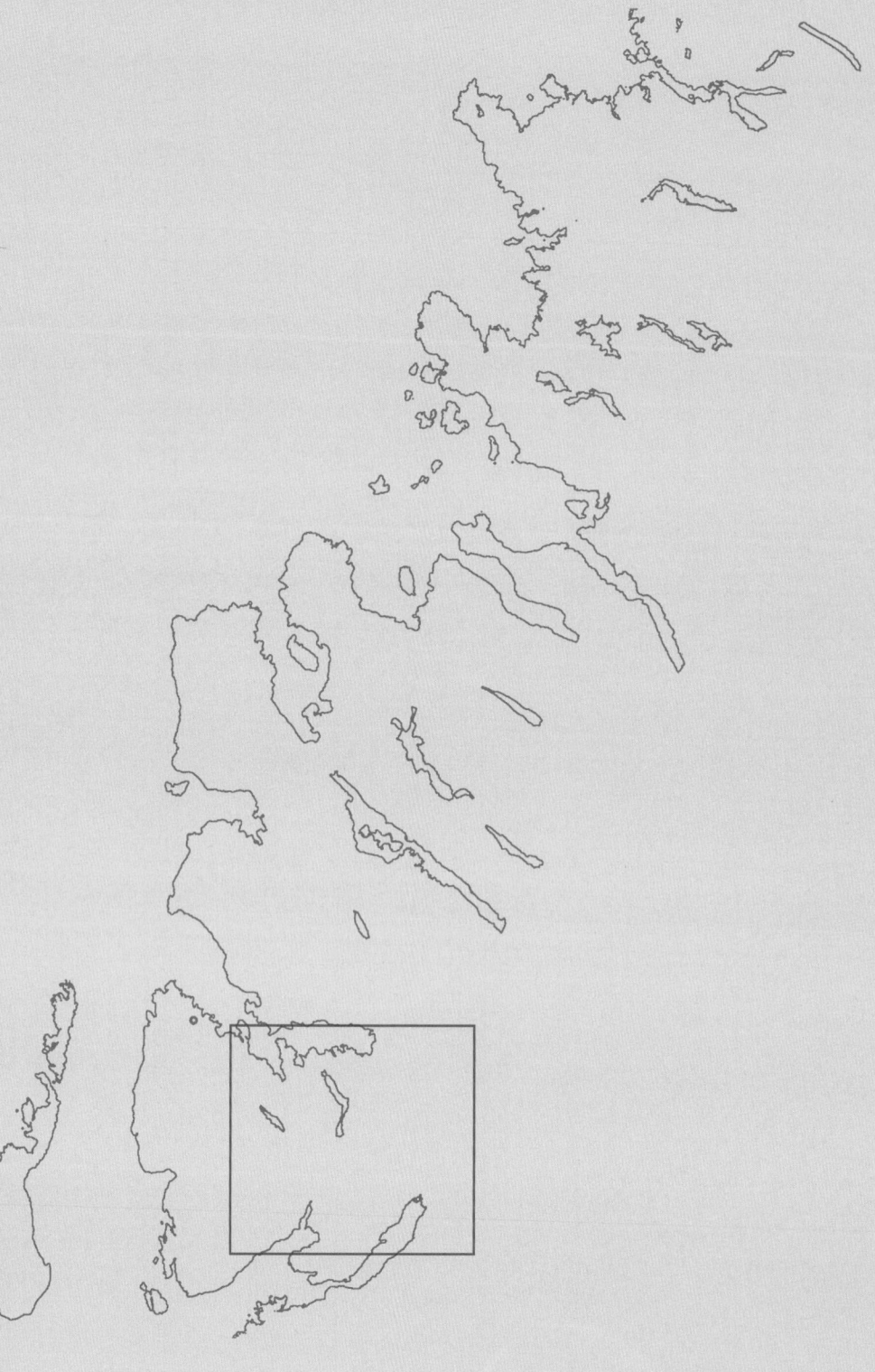

Beinn Liath Mhòr looking towards Beinn Damph

RIGHT: Sastrugi and the Coulin Forest
OPPOSITE: An Ruadh-Stac

LEFT: Rapid below Maol Chean-Dearg

PREVIOUS: Loch an Eoin and Maol Chean-Dearg from Beinn Na h-Eaglaise

Allt Coire Roill and Beinn Damph

Sgorr Ruadh and Fuar Tholl from Beinn Alligin

Sgùrr Dubh and the Coulin Forest from Creag na Rianaich

COULIN ADVENTURES

Although safety is a factor in the winter months, walking alone in the summer doesn't concern me, I just prefer company. When everything goes to plan and that perfect combination of light and location presents itself I love to share my excitement. When a hike starts to unravel it's equally nice to have a friend at your side and complain about what a cruel world it is! Shared experiences make for great friendships and there is always plenty of time to talk when you are walking.

I've been tested a few times out on my own and wished for company, but I've only once experienced near disaster. My first visit to the Coulin Forest came one May three years ago and having visited the other areas of the Northwest several times I was looking forward to visiting a completely new set of mountains. I picked Sgorr Ruadh for my first camp expecting wonderful views out to the west and across to the neighbouring peak of Maol Chean-Dearg.

The forecast was excellent. High broken clouds and light summit winds were predicted for the evening and the following morning - I had little to worry about. The seven kilometer route up from Glen Torridon was gradual and well pathed and I made quick work of it. The final ascent, although steep, was straightforward with expansive views unravelling before me. A strong breeze gave cause to question the forecast, but I set up my tent close to the summit safe in the knowledge that it couldn't be THAT bad.

As sunset approached the cloud level lowered and the wind accelerated. Sunrays burst through the cloud and sent golden beams dancing across the landscape (page 35). I busied myself taking photos, struggling to stabilise the camera in the wind. Eventually the cloud swallowed the mountain and I returned to my tent to escape from the weather.

Half an hour later the tent was being battered by gale force winds. With the poles little match for the power of the passing gusts I did my best to support them, but without success. A snapping sound preceded a gradual collapse of the tent. I began hurriedly packing away, conscious that with one pole down others would surely follow. Somehow I managed to get everything back into my pack without losing any of my kit to the gale. With the light failing I raced off the hill and back down to my car cursing the weather forecast under my breath!

In April of 2018 I returned a veteran of the area having hiked all but Maol Chean-Dearg and An Ruadh-Stac. This time Harsharn, a fell runner and regular companion, joined me. A cursory glance at the forecast was all it took to abandon our camping plans. The Mountain Weather Information Service leaves you in no doubt when bad weather is coming. With 60mph gusts forecast there was the rather emphatic note: 'Walking tortuous'. We decided to stay in a bothy.

Bothies are small mountain huts dotted all over the Highlands, generally with no facilities and accessible only by foot. Originally kept for farm labourers or estate workers, these small buildings have been repurposed by the Mountain Bothies Association as refuges for hill walkers. Free to use, they rely on a simple common sense code of ethics, the Bothy Code, and are maintained almost entirely by volunteers.

In the evening, in preparation to hike the following day, we headed to the Coire Fionnaraich Bothy, a typical example. Downstairs are two basic living rooms and upstairs two bedrooms. Wood panelled walls keep any warmth in and the drafts out, but it is the multifuel stove downstairs that forms the heart of the Bothy.

With only a few miles to walk we carried in plenty of coal and more food than we could eat. There we met Doug and Joshua, a father and son team from the States who were part way through hiking the Cape Wrath Trail, a 205 mile hike connecting Fort William to the most north-westerly point of mainland Britain. With the fire lit and the room quickly warming up we spent the evening sharing hiking stories and our much tastier food!

The following morning it was raining and our bothy companions set out early, fully clad in their waterproofs. By 11am the sun was out and I was regretting not starting sooner. The visibility was perfect and billowing cumulonimbus drifted overhead, occasionally bringing a shower. The winter grasses on the hills were rich oranges and yellows from the morning's rainfall. This was a rare day to capture the atmosphere and drama of the Scottish Highlands.

We set a solid pace up from the bothy stopping only briefly to top up our water. Hiking with a daypack is significantly easier than with a rucksack full of camping kit! Our route up the leeward side of the mountain was mostly sheltered from the westerly winds, but occasional gusts whipped around, a warning of what awaited us above. We reached Bealach a' Choire Ghairbh and came face to face with the gale.

I left my pack at the pass, taking just my camera, and we headed on up An Ruadh-Stac, once again finding ourselves protected from the worst of the wind. The further we climbed the more promising the clouds became.

At the top and exposed to a ferocious wind we were presented with a plumetting view over Loch Coire an Ruadh-Stac and across to the mountains beyond. A beautiful patchwork of light stretched across the landscape, moving rapidly. Out to the west a heavy downpour approached - surely this was the drama I was after. I knew that once the rain had passed through I would have a breif window to capture the dynamic conditions. I just had to wait it out. Stashed in my pack were my waterproofs...three hundred meters down the mountain.

For a while I considered my options, but there was only one acceptable choice. It was better to get wet, than miss something special. 'Go big or go home'. As the rain hit, Harsharn sat smugly in his waterproofs. I hunkered down.

The rain was intense but the conditions that followed vindicated my decision. Malevolent clouds hovered over the mountains of Torridon, a wall of rain obscured the peaks. Fierce sunshine broke through the clouds behind me, lighting up the flanks of Beinn Damph and Maol Chean-Dearg. A rainbow arced over Sgorr Ruadh to the west - the crowning glory of a spectacular scene (next page).

The day wasn't done yet. We dashed back down the mountain, I collected my pack, and we headed straight back up Maol Chean-Dearg stopping briefly to photograph another rainbow (page 31). Although it scarcely seemed possible the wind had intensified and we both struggled to stay on our feet, occasionally falling as we laughed our way to the summit.

More views, more rain, more wind, more drama - Scottish weather at its finest (page 30).

Maol Chean-Dearg from
An Ruadh-Stac

ABOVE: Meall nan Ceapairean from Maol Chean-Dearg
OPPOSITE: Loch an Eoin and Liathach from Maol Chean-Dearg

LEFT: A' Phoit and A' Chioch, Beinn Bhàn
PREVIOUS: The Coulin Forest from Liathach

Maol Chean-Dearg from Sgorr Ruadh

Maol Chean-Dearg and Sgorr Ruadh from Fuar Tholl

The Beinn Damph and Coulin Forests
from Beinn Bhàn

TORRIDON

From Loch Torridon to Loch Maree

Torridon is a mecca for Scottish hill walking. Its three Munros, Beinn Alligin, Liathach, and Beinn Eighe, are regularly listed amongst hikers' favourite hills, and with just cause. Each of these mountains can be visited in an exciting loop with plenty of exposure and inspiring views of the surrounding peaks and lochans below. Climbing opportunities are equally well known; multiple winter and summer routes snake their way through the endless north facing crags. To the north of 'The Big Three' more placid hills offer spectacular views into the mountains to the south and plenty of solitude. Of these, Baosbheinn is a personal highlight. Its extended ridge offers numerous photographic opportunities.

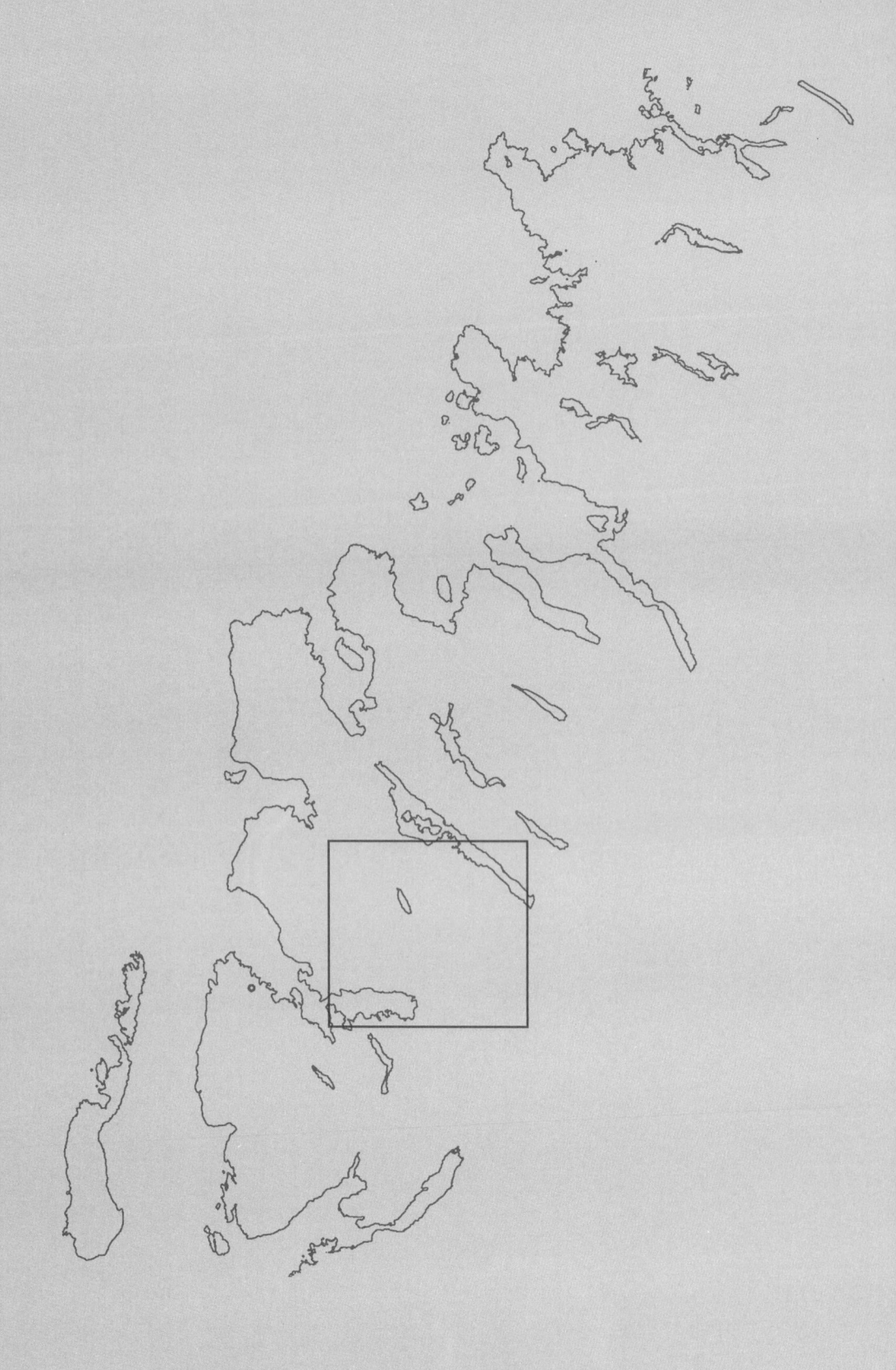

TORRIDON

LOCH MAREE

FLOWERDALE

Baosbheinn

Loch a' Bhealaich

Loch na h-Oidhche

Beinn an Eoin

Srath Lungard

Beinn a' Chearcaill

River Grudie

Meall a' Ghuithais

Sgurr Mor

Beinn Dearg

Coire Mhic Fhearchair

Ruadh-stac Mor

Tom na Gruigach

BEINN ALLIGIN

Sail Mor

Spidean Coire nan Clach

TORRIDON

BEINN EIGHE

Coire na Caime

Mullach an Rathain

Spidean a' Coire Leith

LIATHACH

Creag na Rianaich

Loch Clair

LOCH TORRIDON

Sgurr Dubh

COULIN FOREST

LIATHACH IN WINTER

It's always been my belief that artists should have a desire to differentiate themselves. For many photographers individuality develops simply as a byproduct of practicing their craft. Subtleties surrounding the selection of scenes, prefered lighting conditions, compositional choices and variations in processing can all lead to distinctive styles emerging. In some cases a photographer can be identified from a single image based solely on these distinctions. However, my personal preference for illustrative, realistic imagery in a 'grand scenic' style leaves me a little more restricted. I try to let the landscape do the talking, but in doing so talk a little less myself.

Shortly after taking up photography I found myself trying to capture scenes that were as fresh as possible to the viewing public and landscape photographers alike. I came to realise that photographing popular locations was deeply unrewarding to me personally. Photographing clichés felt redundant.

As the digital revolution progressed around 2010, Dartmoor, my initial proving ground, became more photographed and that ultimately drove me further afield in the search of images that were increasingly hard to achieve. What started with overnight trips on the middle of the Moor lead to summit camps in the mountains, and eventually to multi-week backpacking trips covering many miles in Norway, Iceland, Greenland and South Africa.

Difficulty is not a quality that can be admired in a photograph. Effort cannot be seen in an image, nor can a photograph express the stories leading up to that moment. Yet difficulty, within reason, is a primary motivating factor behind my approach. My photographs give me creative enjoyment of course, and I can appreciate them visually, but their importance to me is as much tied to the experience as it is to the result. The more I have struggled to capture an image, the more I value it and that worth is largely divorced from the visual standard I managed to achieve. Hiking helps me to differentiate my work, but it also helps me to value it.

When it came to planning this book there were a few peaks that were essential inclusions and chief among them was the Torridonian peak of Liathach - Greyish One. I was drawn to the idea of a winter camp next to the iconic Am Fasarinen Pinnacles which make up the most technical section of the ridge traverse of Liathach. It would be an opportunity to capture an epic winter scene whilst at the same time banishing the demons of An Teallach (page 84).

Liathach's knife edge ridge caused me to question whether a camp at the top would be possible at all, so on a baking August afternoon I escaped the midges in the valley below to recce the route with my friend Geoff. This turned into one of the most enjoyable hikes of my life with plenty of exposure, enjoyable scrambling and, above all, views that genuinely amazed me - something increasingly hard to achieve after adventures in more extreme locales. For Geoff the hike also encompassed his first two Munros, which was quite the introduction! Most importantly I found a spot for a two man tent, right next to the Pinnacles. The winter mission was on.

Sadly patience comes into play more often than I would like and the winter of '16-'17 proved to be mild and wet. As every cold snap hit the Highlands I grew hopeful of something more sustained, but as the months passed I grew more frustrated. It would have to wait for another year.

A January snowfall in 2018 and a dose of recklessness was all it took to get me back up to Torridon ascending the southern face of Liathach, this time with Twitter friend Sven Soel in tow. It's fair to say that we were both a little out of our depth, both literally and metaphorically. Our optimism following fast progress on the largely snow-free lower slopes soon gave way to a sense of hopelessness as we hit thick snow and realised the route might be beyond us.

The steepest section starts 200m below the main ridge where, in summer, you can pick your way between outcrops and follow the feint semblance of a path. It was here that we found the greatest difficulty. With the snow knee deep we couldn't lift our legs free and instead we were forced to plough our way through, taking it in turns to break trail whilst at the same time remaining all too aware of the slope falling away beneath us. It is in these circumstances that having a hiking partner makes all the difference. Part way up the snow slope Sven was ready to give up, whilst I was determined to carry on. Later the roles were reversed when Sven's positivity kept me going. Between us we found the confidence to complete the ascent.

We reached the easterly Munro of Spidean a' Choire Lèith a full two hours behind my estimated arrival time and only moments before a rather disappointing sunset. My thoughts drifted towards the camp spot that I had discovered 18 months previously, my confidence wavered as I looked down at the ridge - How could there possibly be room for a tent there?

A tetchy descent towards the Pinnacles brought us to my marked GPS spot which, to my relief, was just big enough for the two of us and almost perfectly flat. As darkness arrived, an overhead high pressure system brought with it crystal clear skies and near windless conditions perfect for capturing the view under the light of the half-moon (opposite and page 4). The temperature dropped to -7°C overnight but inside the tent Sven's thermometer read a balmy 3°C. We slept soundly.

Getting up at dawn in the Scottish mid-winter is a joy. Sunrises at 9am are a real luxury for landscape photographers accustomed to early summer starts. I peaked my head out of the tent as the first light appeared on the horizon and felt excitement sweep over me - I was going to achieve my goal.

The views across the Pinnacles towards Mullach An Rathain, Liathach's westerly Munro, were sensational so there was little reason to move from camp. As the light grew I set my camera up to capture a wide panorama (next page) and we watched on as the twilight unfolded over the vast landscape below giving way to a perfect sunrise. Only the soft clack of the shutter broke the silence. Later that morning after fresh coffee and hot porridge we bounded down snow filled gulleys with smiles on our faces.

Although I've spent many wonderful nights camped on summits, such singular experiences are hard to come by. It is to date the finest experience I have had in Scotland punctuated by difficulty and reward in equal measure.

Camping under moonlight on Liathach's main ridge

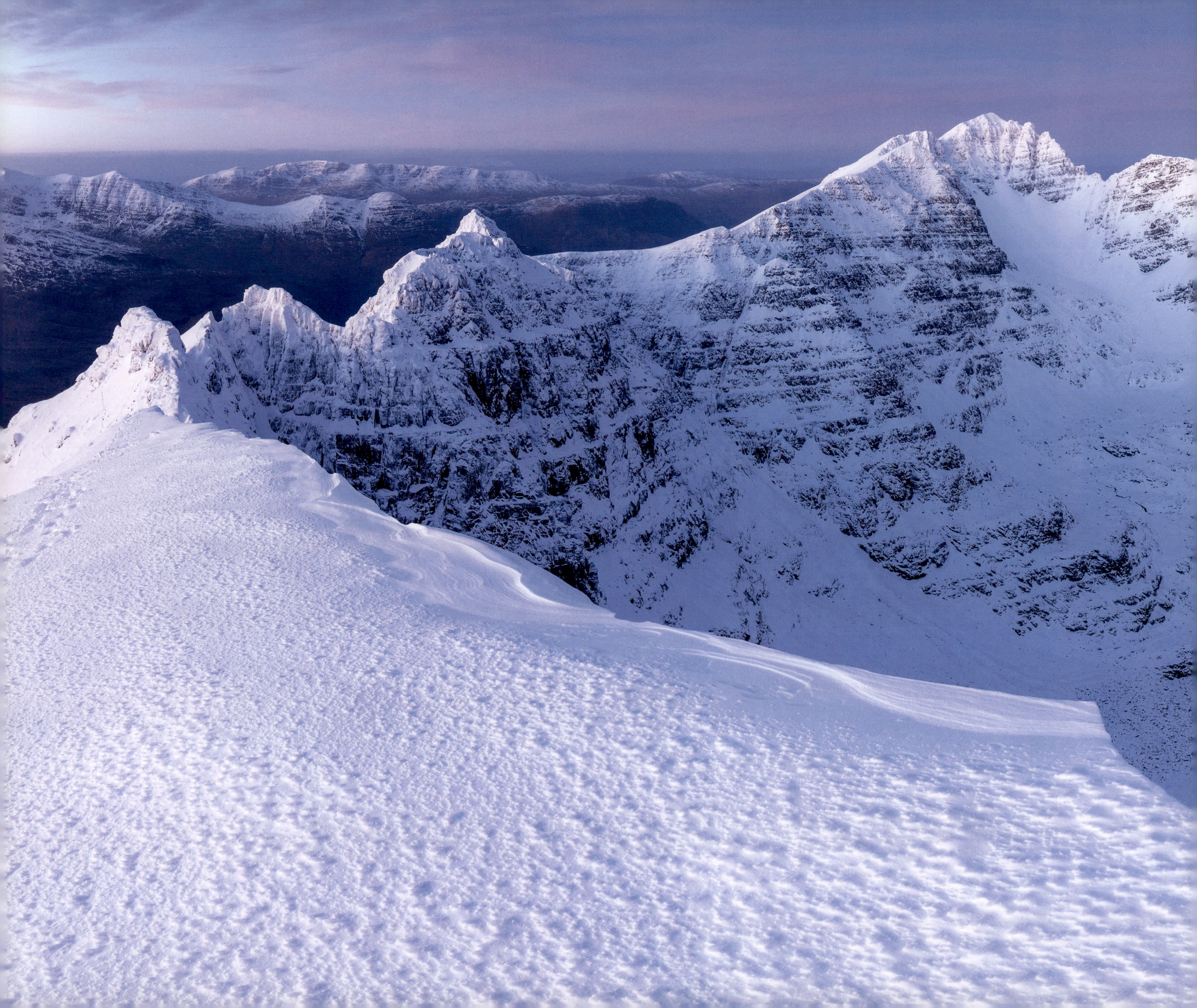

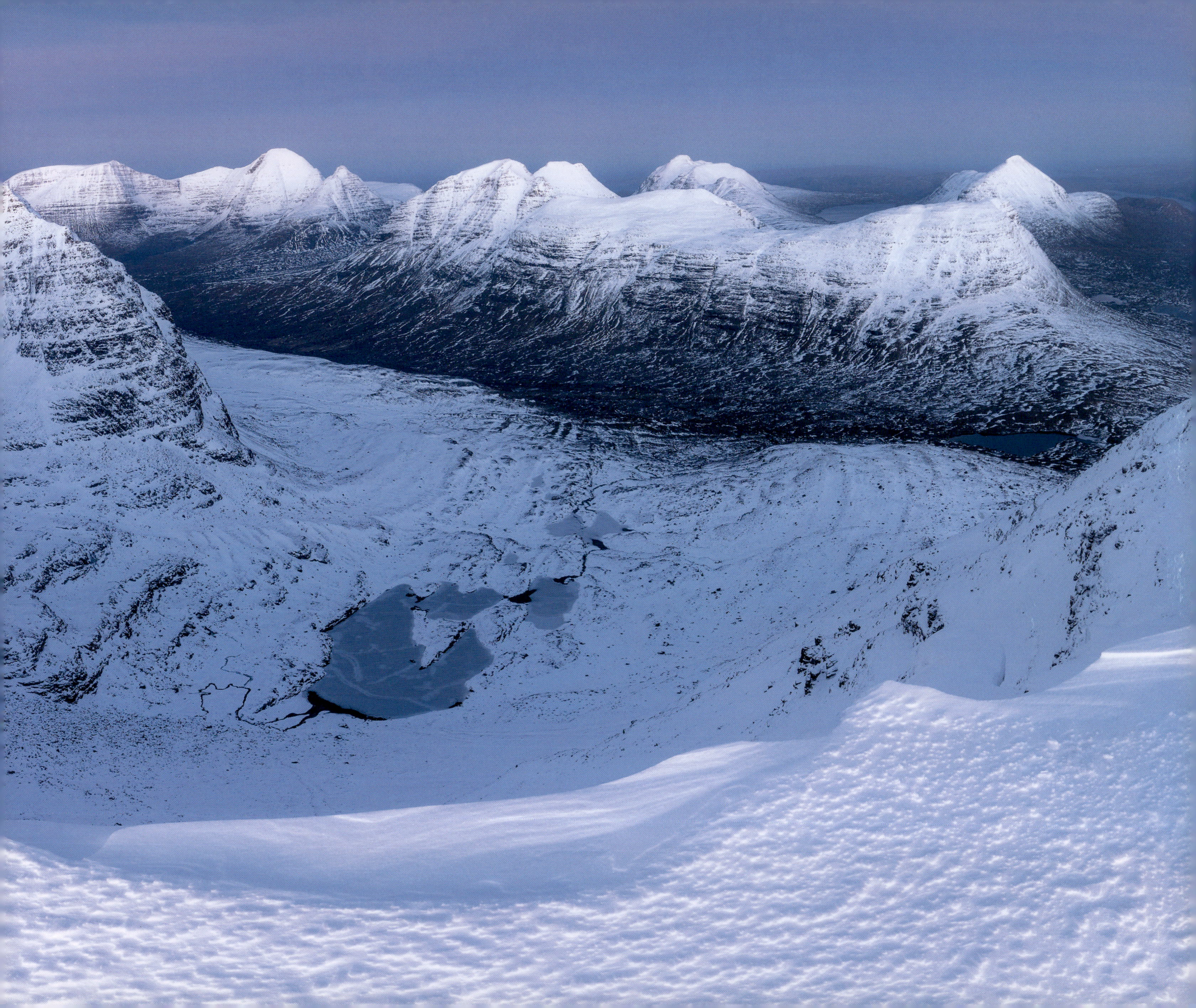

ABOVE: Sgùrr Dubh, Liathach and Beinn Eighe
PREVIOUS: Mullach an Rathain and Coire Na Caime

Am Fasarinen Pinnacles, Liathach
with Beinn Bhàn in the distance

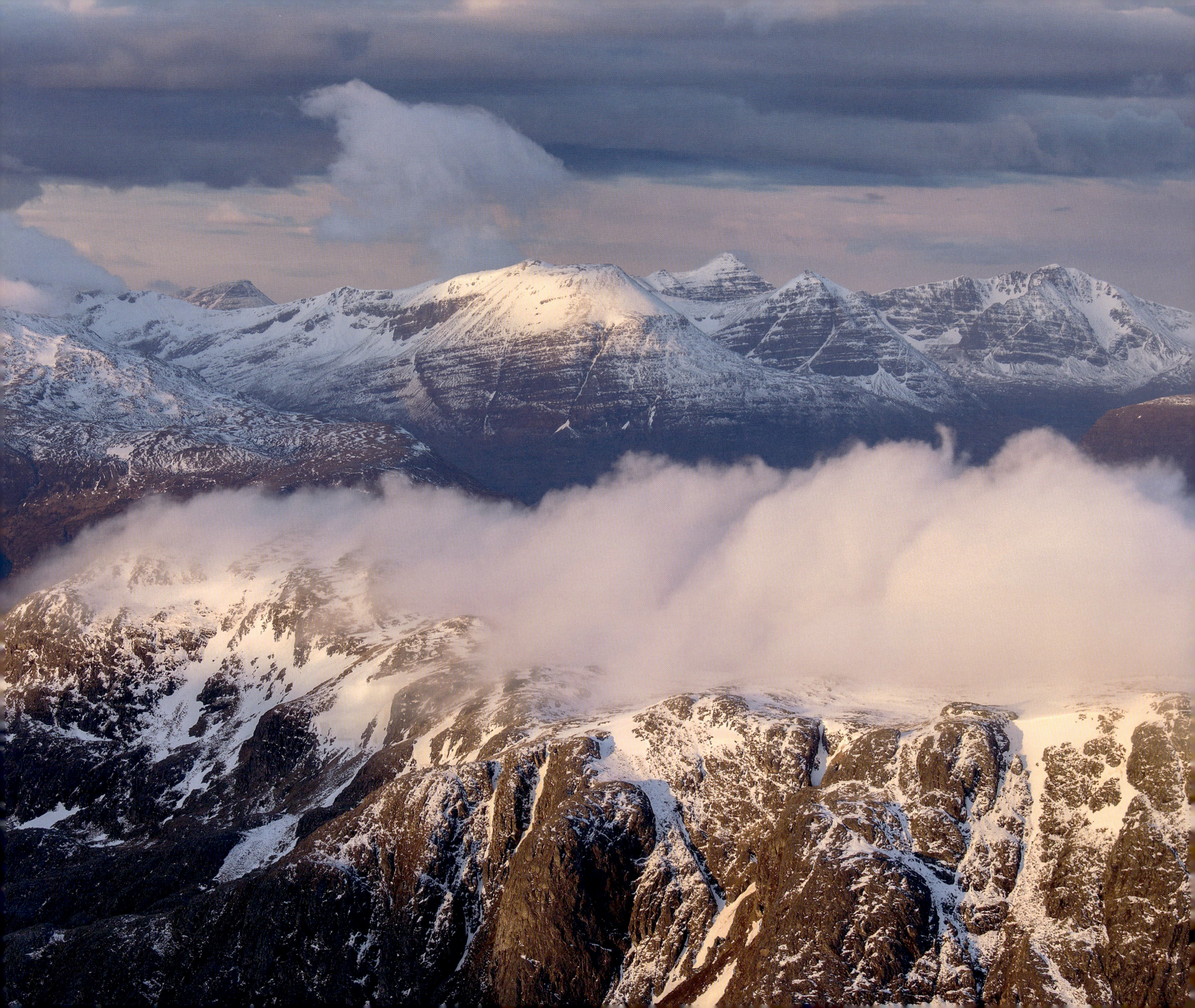

ABOVE: Beinn Alligin from A' Mhaighdean
OPPOSITE: Beinn Eighe and Liathach from A' Mhaighdean

Loch Coire Na Caime,
The Am Fasarinen Pinnacles
and Meall Dearg

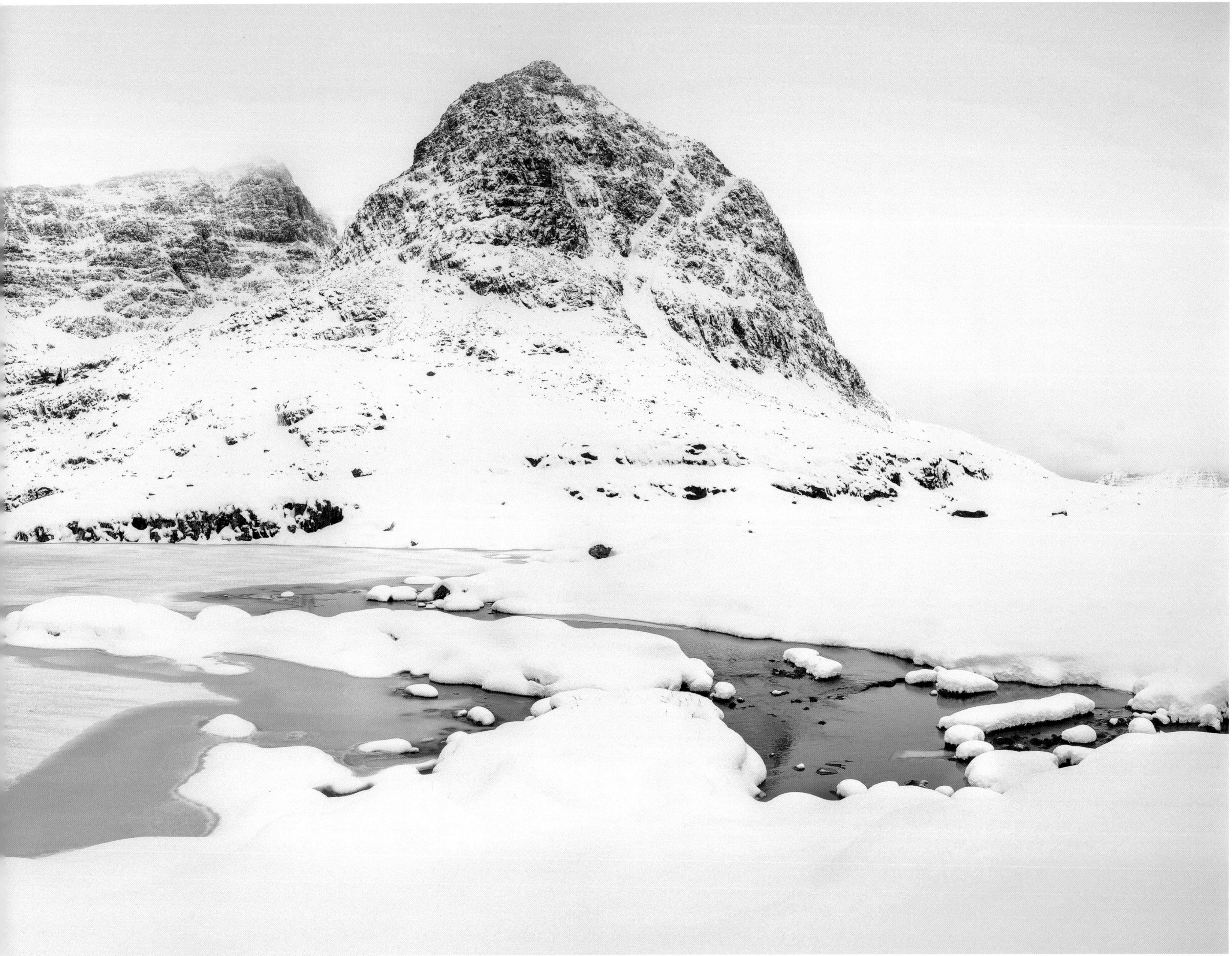

LEFT: Beinn Alligin from Beinn Na h-Eaglaise
OPPOSITE: Beinn Dearg from Coire Na Caime

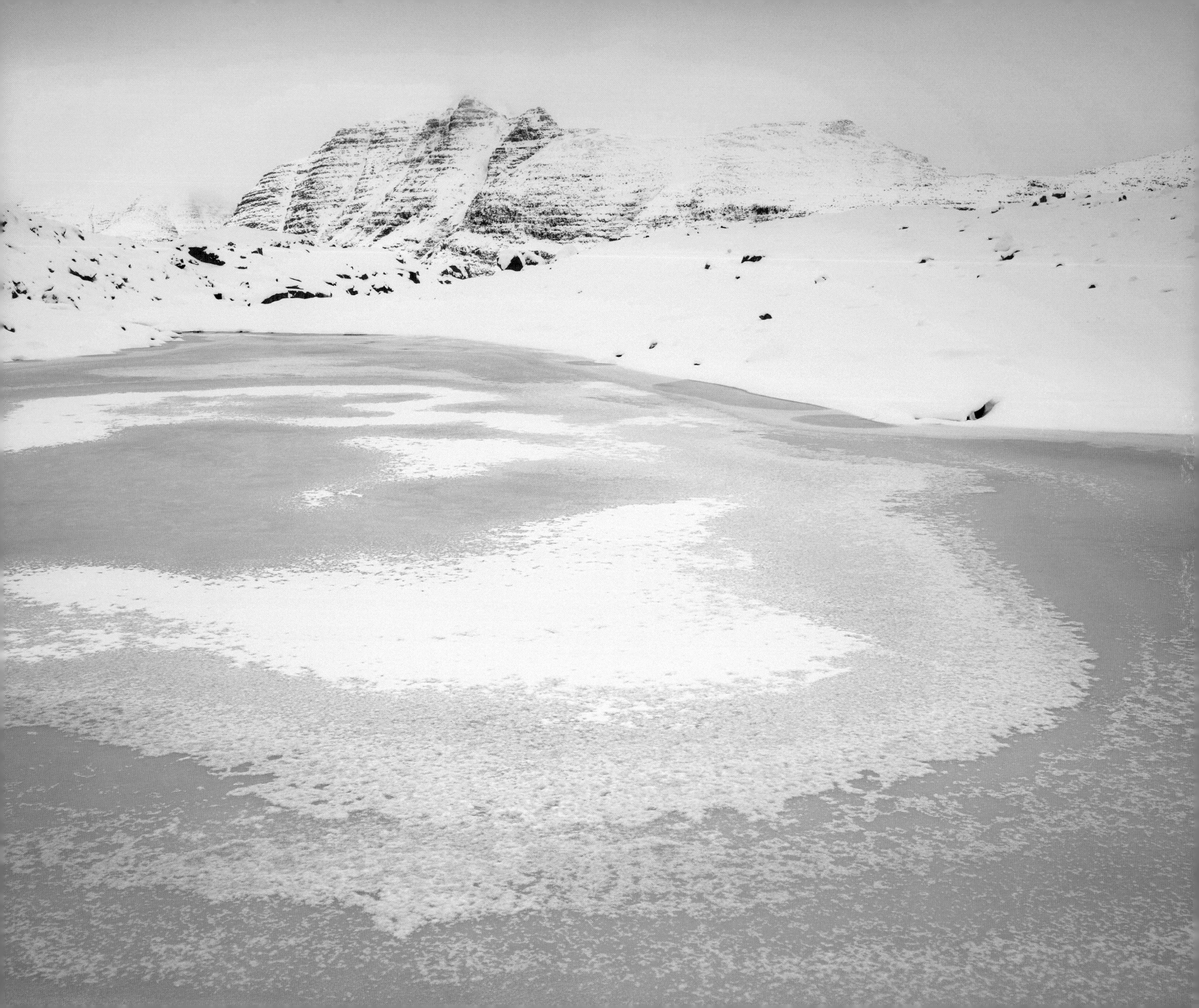

The Horns of Alligin and Beinn Dearg from Sgùrr Mor

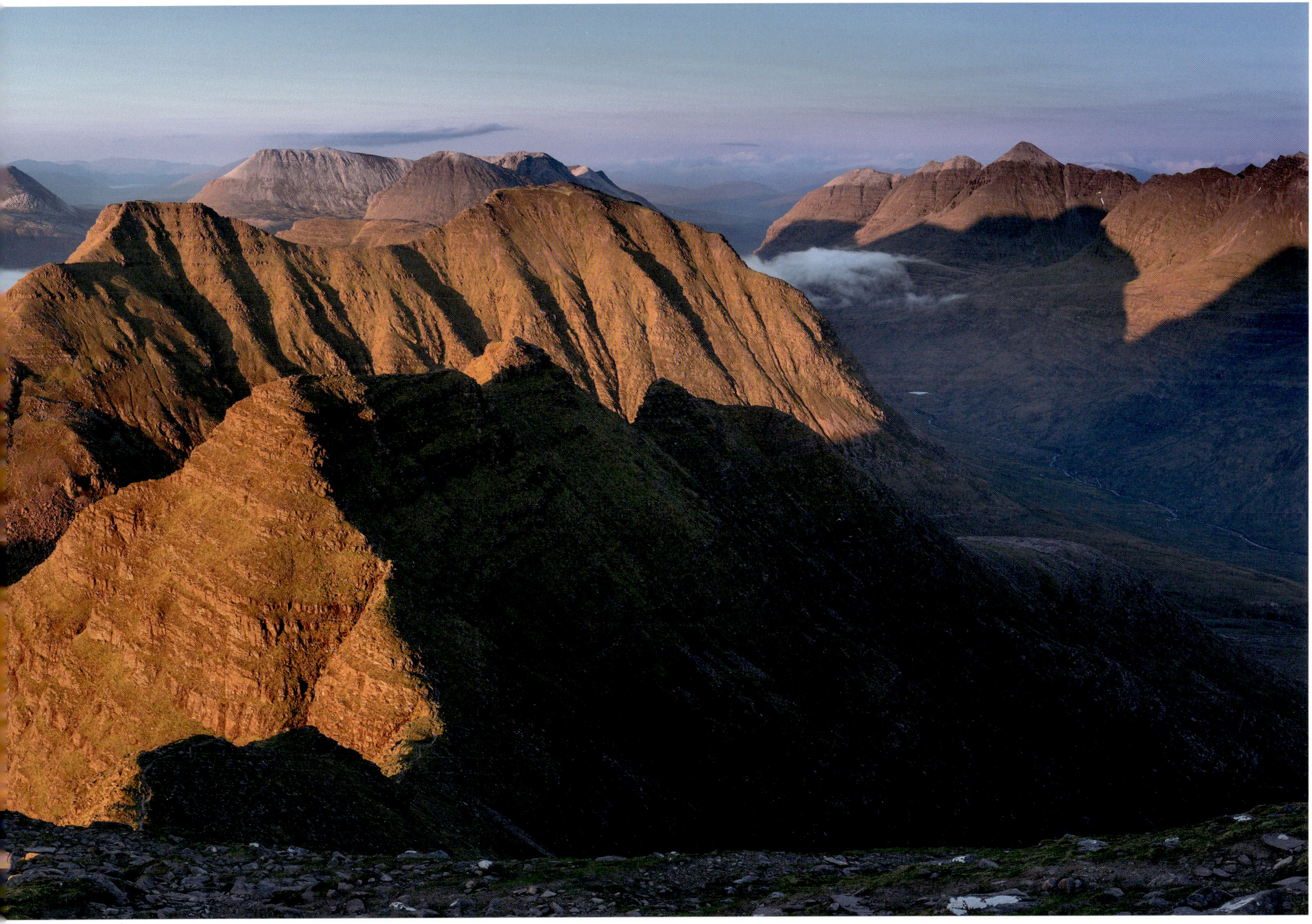

Scots Pine and Loch Clair

Torridonian Sandstone on Beinn Dearg looking towards Liathach

Beinn Dearg and Beinn Alligin
from Baosbheinn

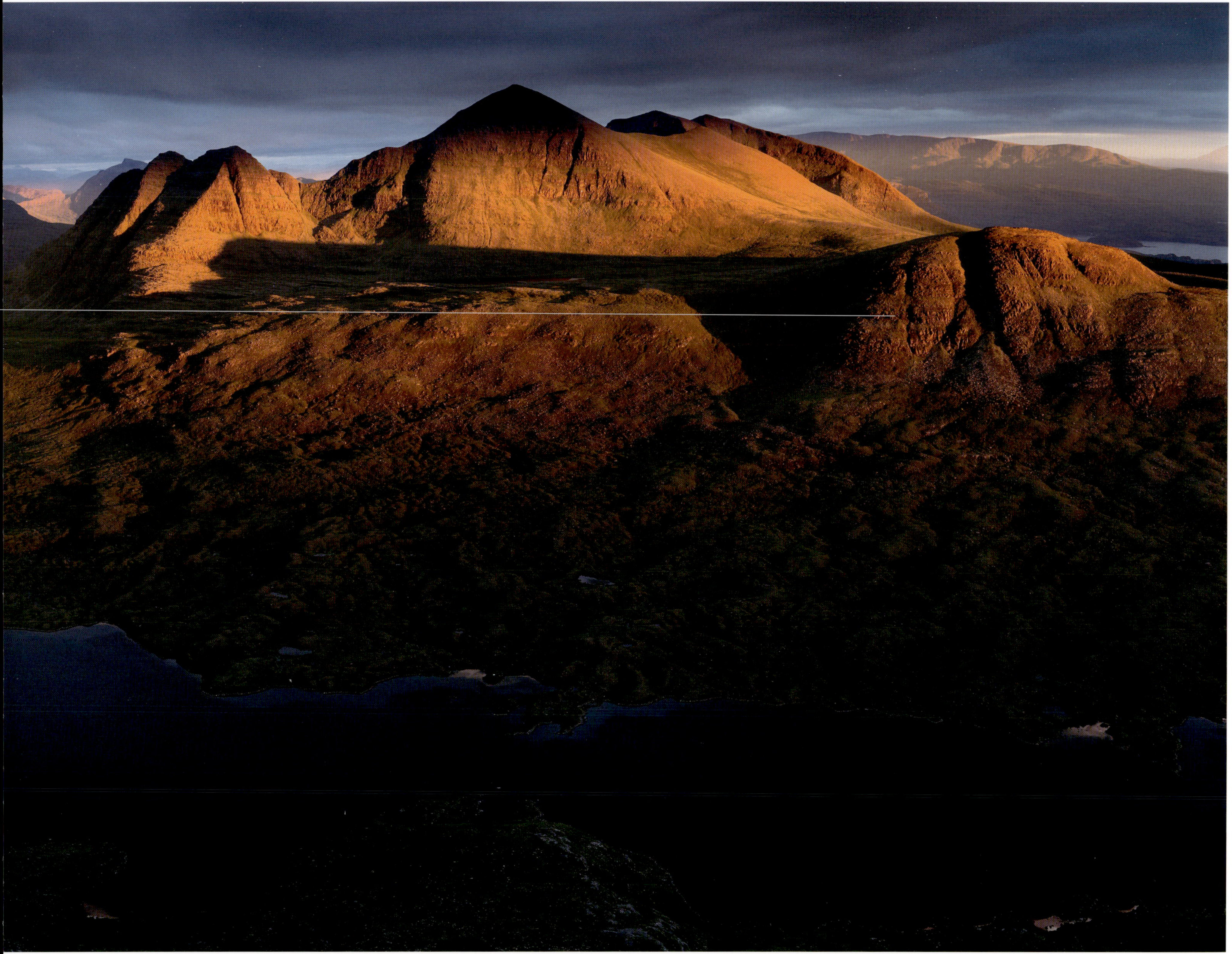

Beinn Dearg and Beinn Alligin

Beinn An Eòin and Beinn Eighe from Baosbheinn

Loch Clair and Sgùrr Dubh

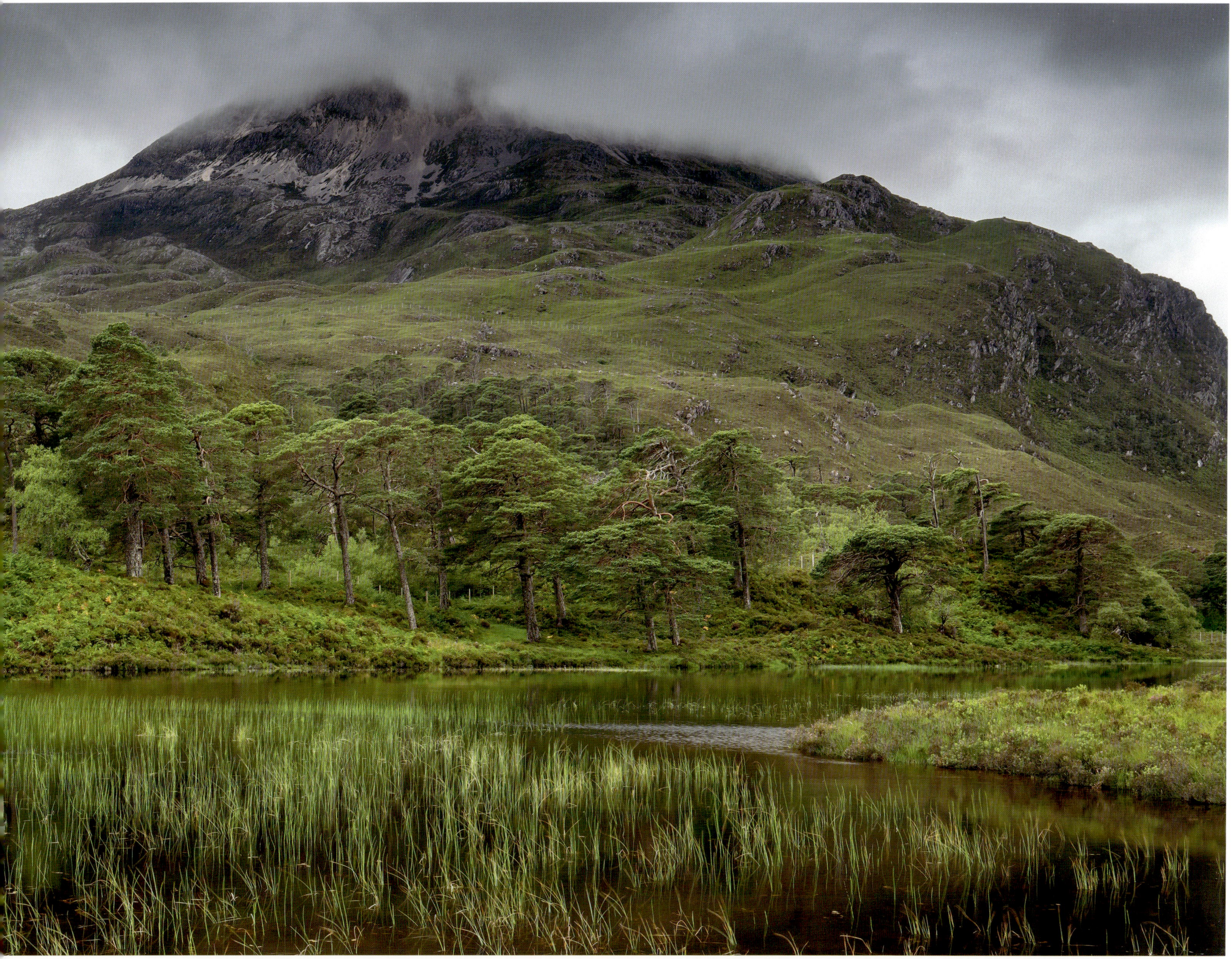

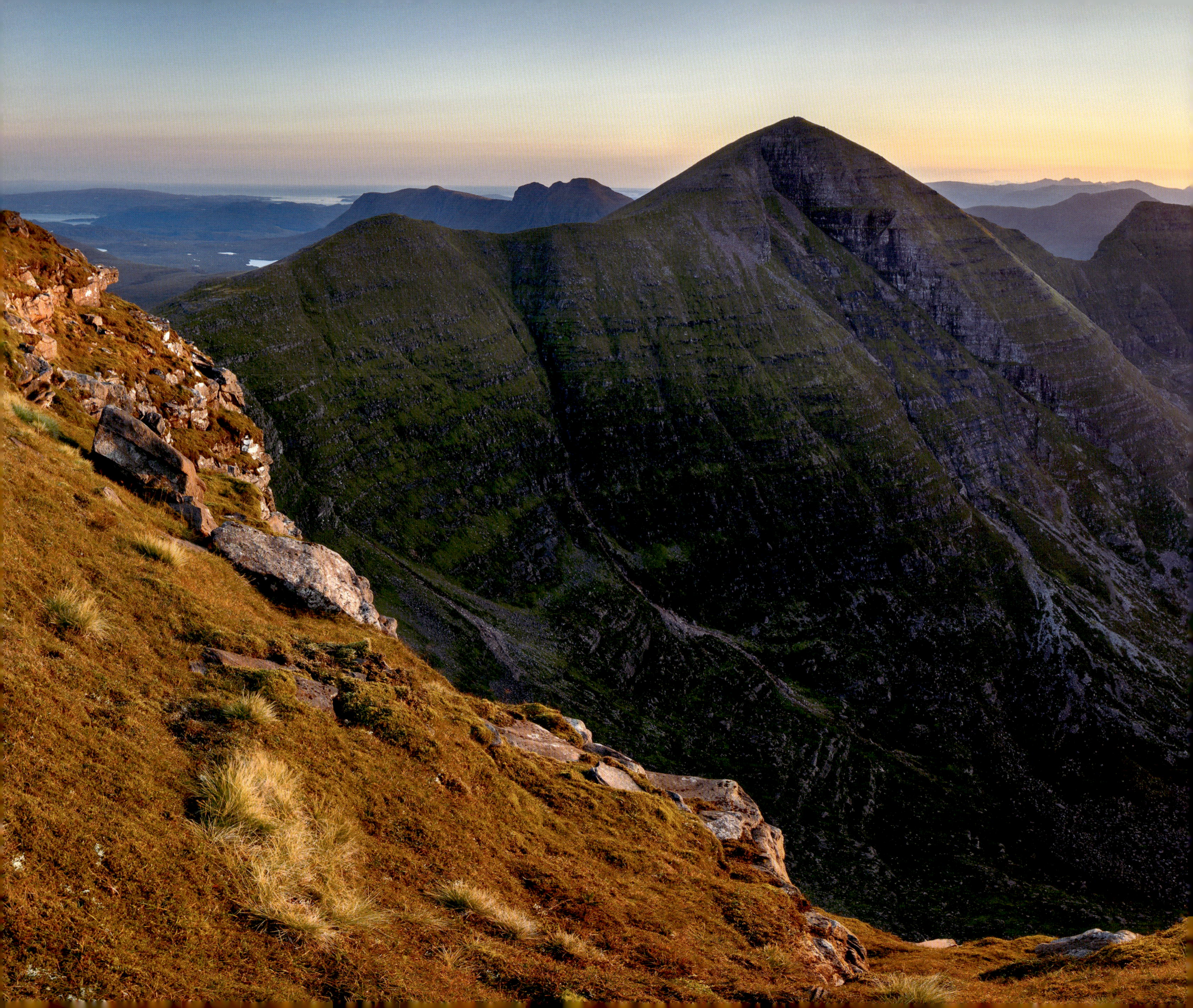

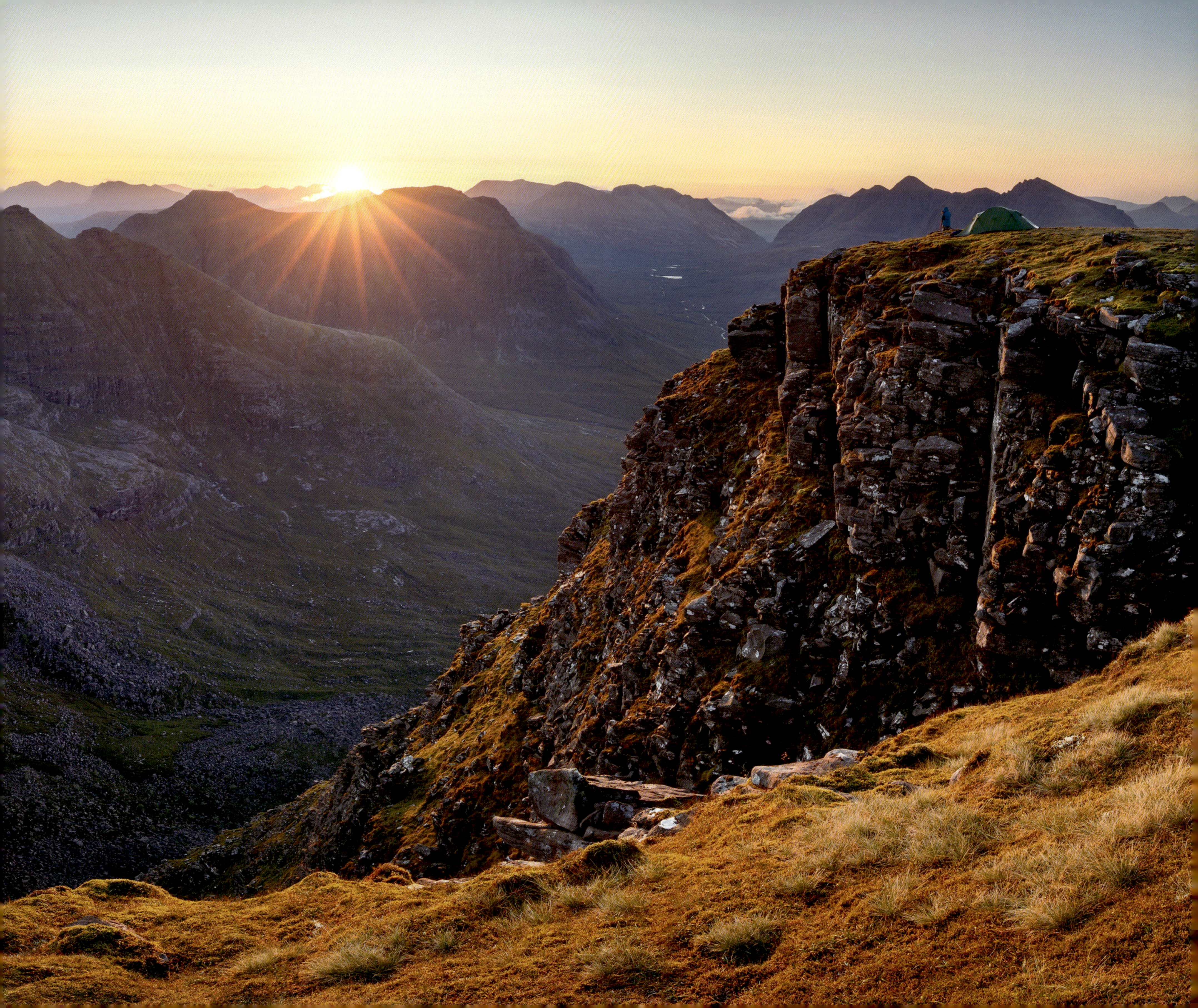

LEFT: Waterfall below the Horns of Alligin

PREVIOUS: Sgùrr Mor and the Torridon Hills from Tom Na Gruigach

The Torridon Hills from Coire Mhic Fhearchair

68

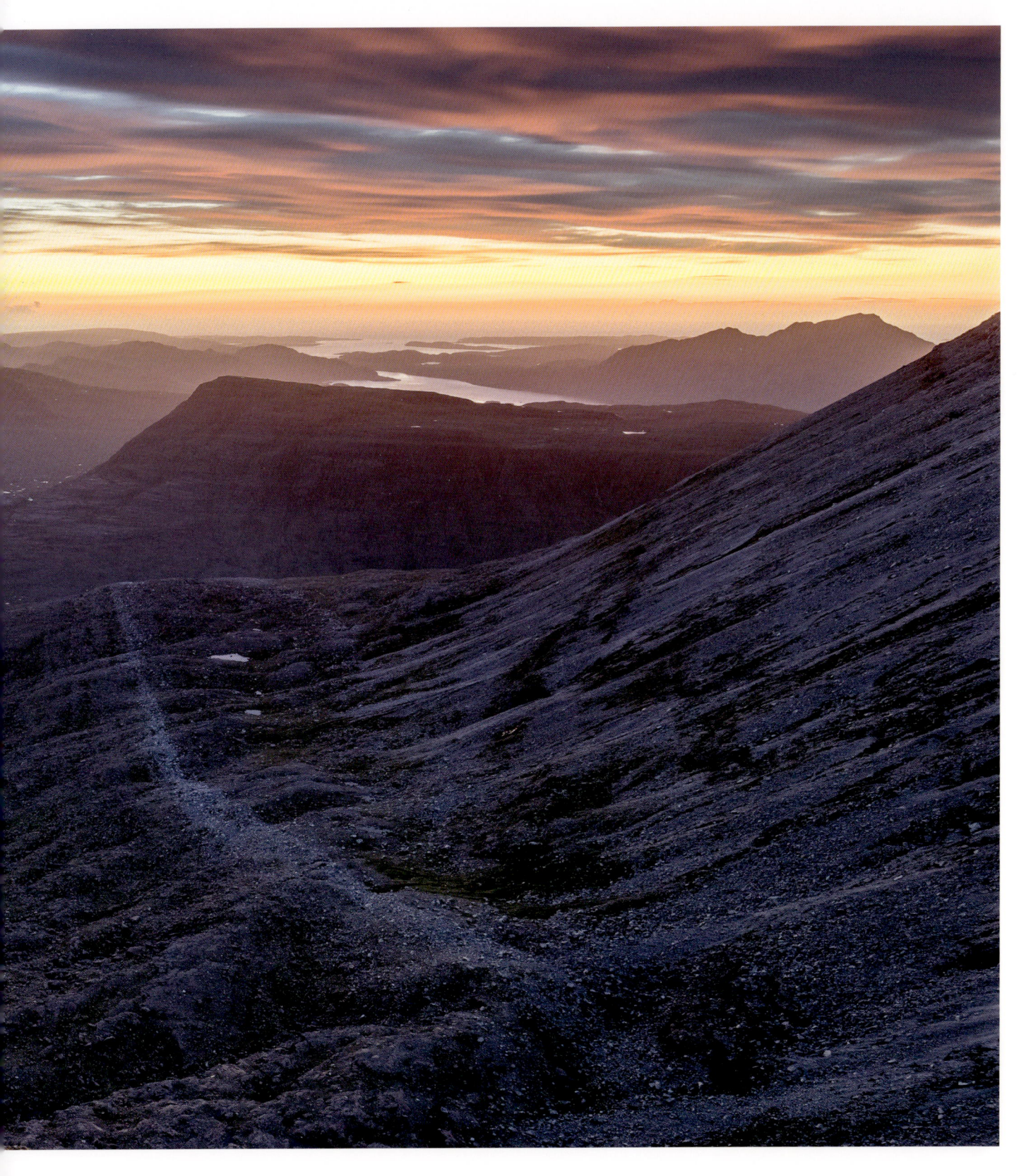

Coire Mhic Fhearchair from
Còinneach Mhòr

Submerged beach and the Horns of Alligin

Torridonian Sandstone on Baosbheinn looking towards Beinn Alligin

THE GREAT WILDERNESS

From Loch Maree to Loch Broom

The Fisherfield Forest, also known as the Great Wilderness, lays claim to being Scotland's most isolated mountain region. A haven for backpackers seeking adventure and solitude, even the easiest mountains provide reasonable difficulty to hillwalkers given their distance from the road. Home to Scotland's most remote Munro, A' Mhaighdean, attempting its peak in a single day is a mammoth task - a 27 mile return route from Poolewe. But the views over Fionn Loch out to the west and over Gorm Loch Mor to the south are unforgettable. An Teallach in the north of the region is another revered peak. Its airy ridge scrambles test even the most confident hiker and in the winter it becomes one of the great Highland challenges. There are many other mountains here worthy of a visit, not least Beinn Tarsuinn with its wide vistas of the Fisherfield 6 and Slioch which dominates Loch Maree to the south.

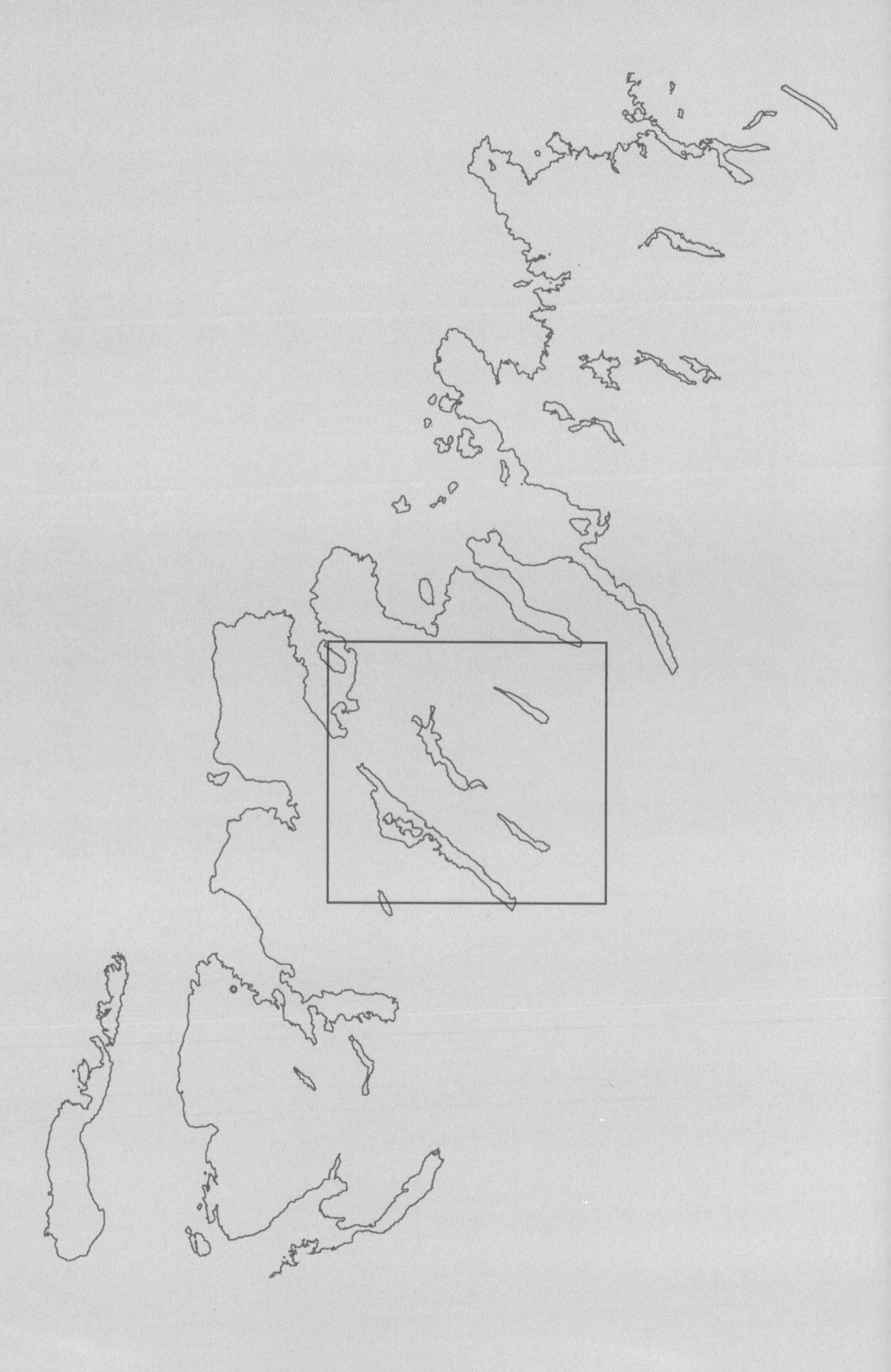

LOCH EWE

AN TEALLACH
Bidean a' Ghlas Thuill
Sgurr Fiona
Loch na Sealga

Poolewe

Fionn Loch   FISHERFIELD

Beinn Dearg Mor

Beinn a' Chlaidheimh

Beinn Airigh Charr

Loch Dubh

Ruadh Stac Mor
A' Mhaighdean

Sgurr Ban

Mullach Coire Mhic Fhearchair

LOCH MAREE   LETTEREWE

Beinn Lair

Islands

Beinn Tarsuinn
Lochan Fada

Slioch

FLOWERDALE

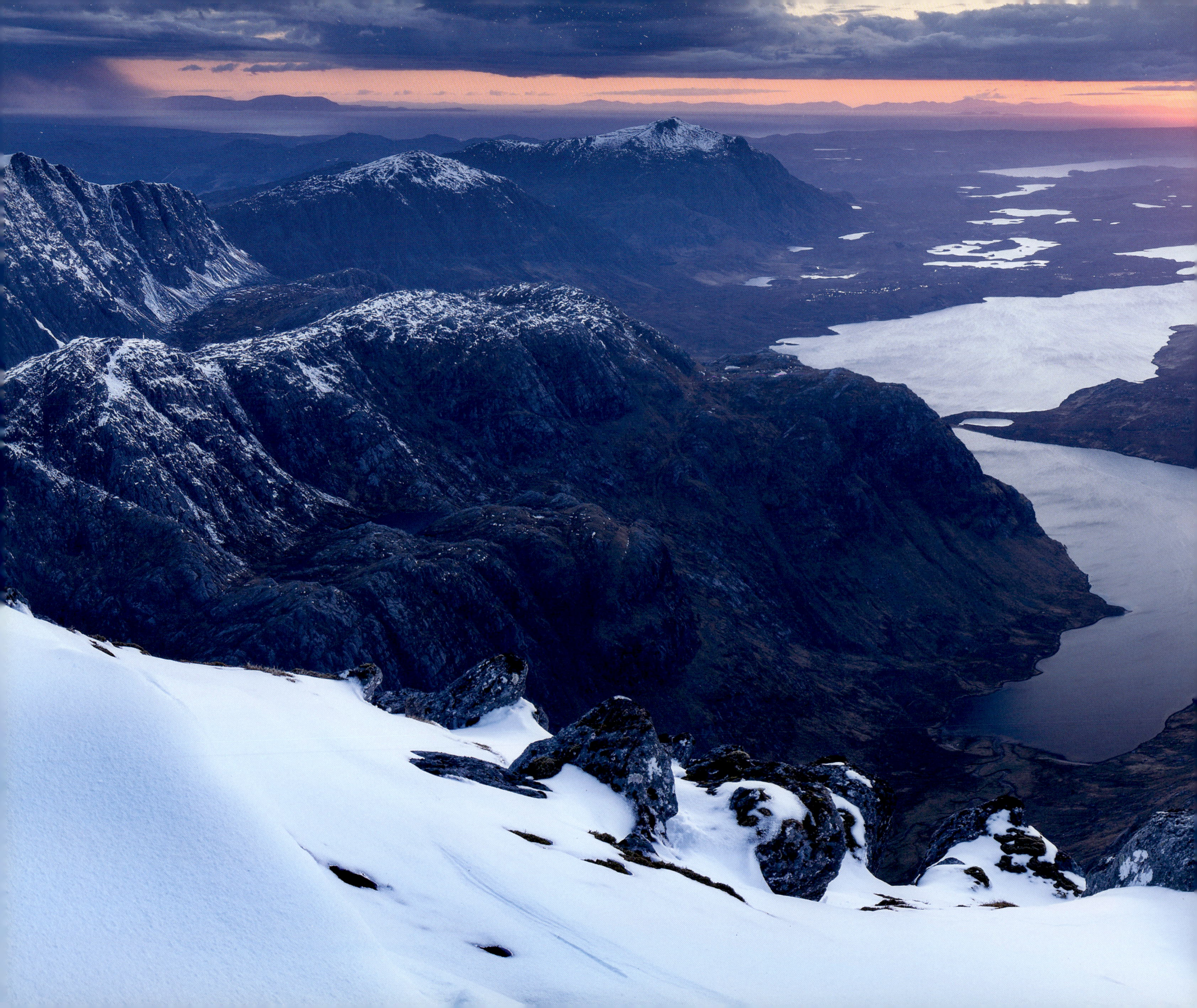

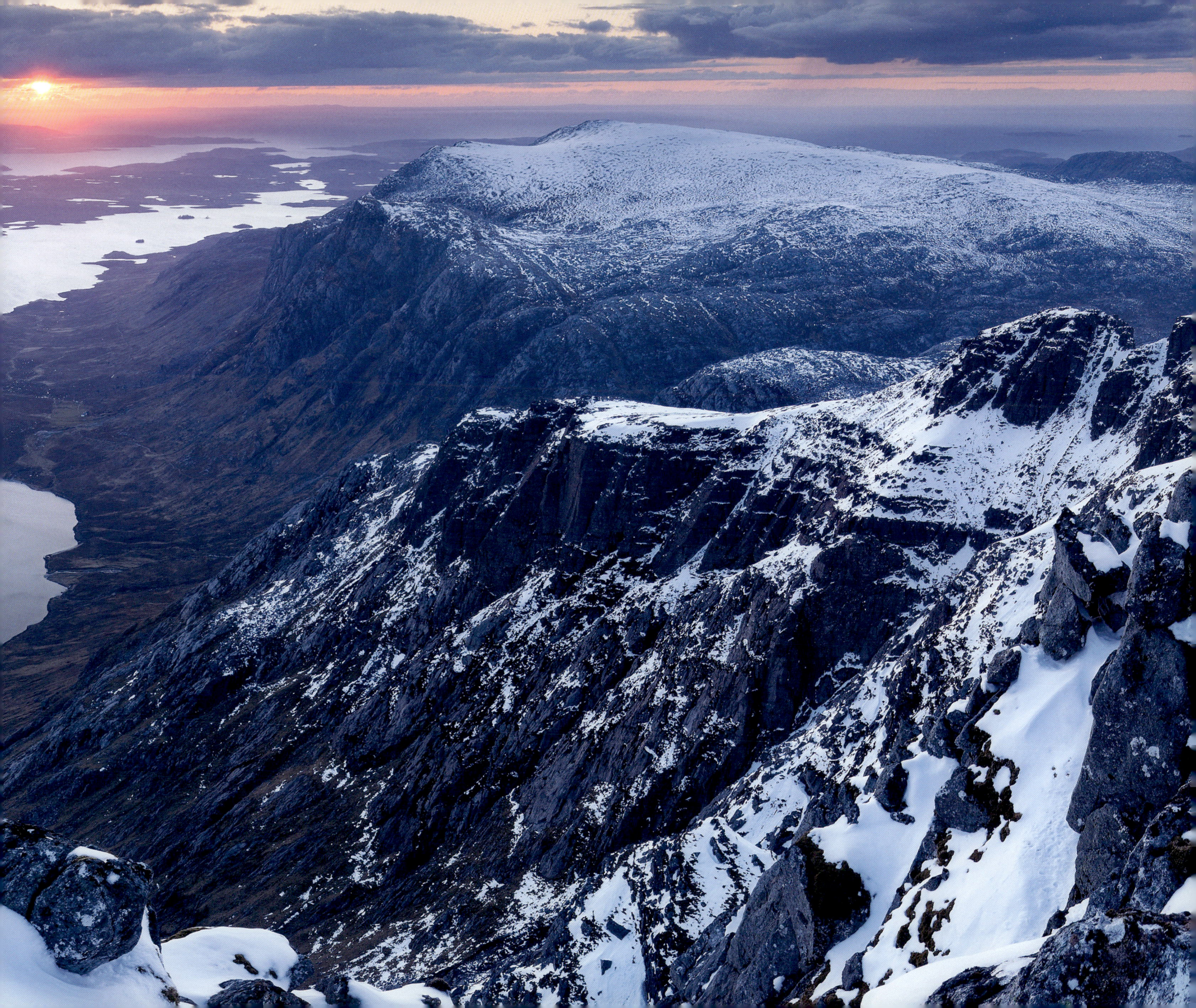

RIGHT: Mullach Coire Mhic Fhearchair

PREVIOUS: Dubh Loch and Fionn Loch from A' Mhaighdean

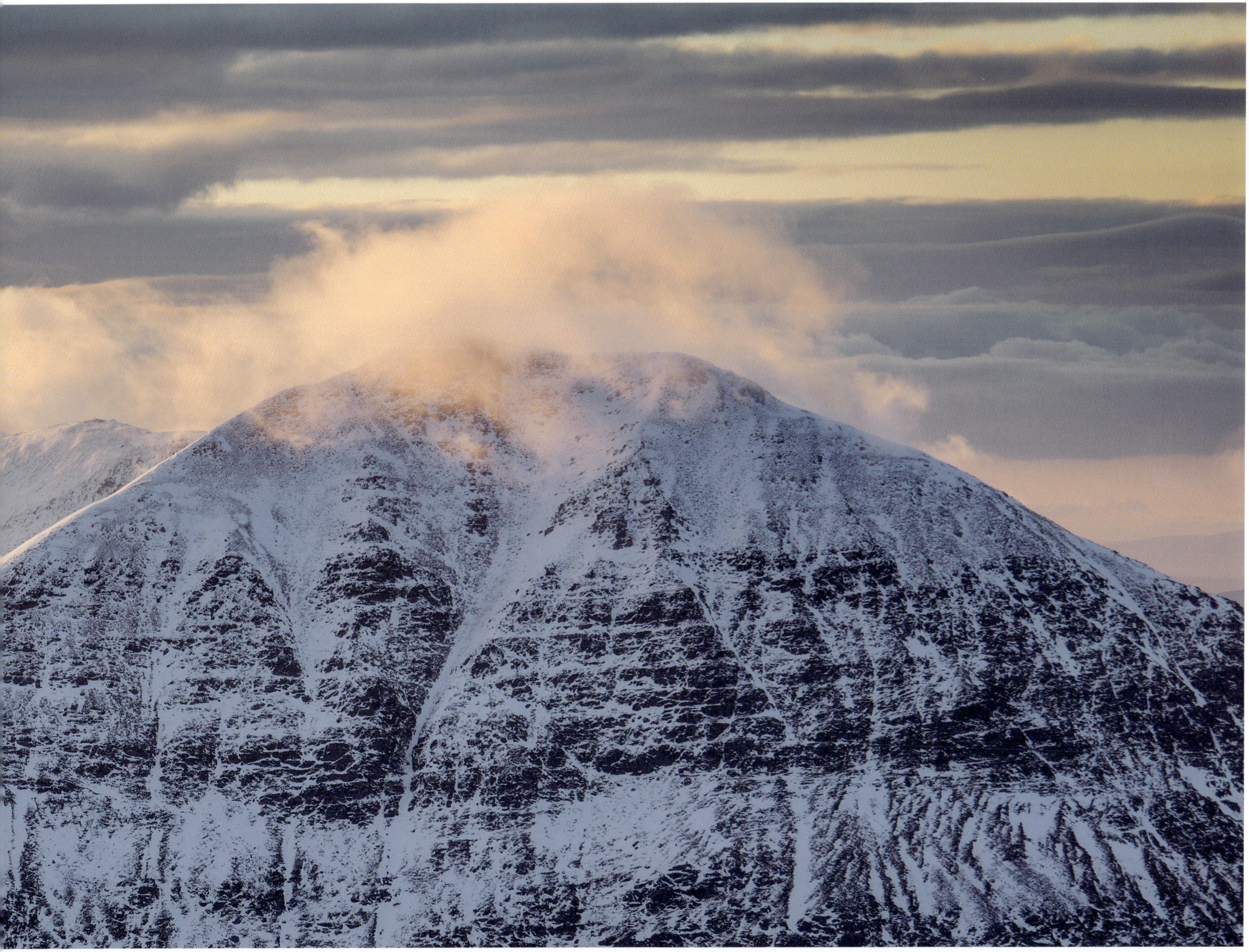

Sgùrr Mor from A' Mhaighdean

Sgùrr Creag an Eich and Sgùrr Fiòna, An Teallach

Snow shower, Loch Maree and Slioch

81

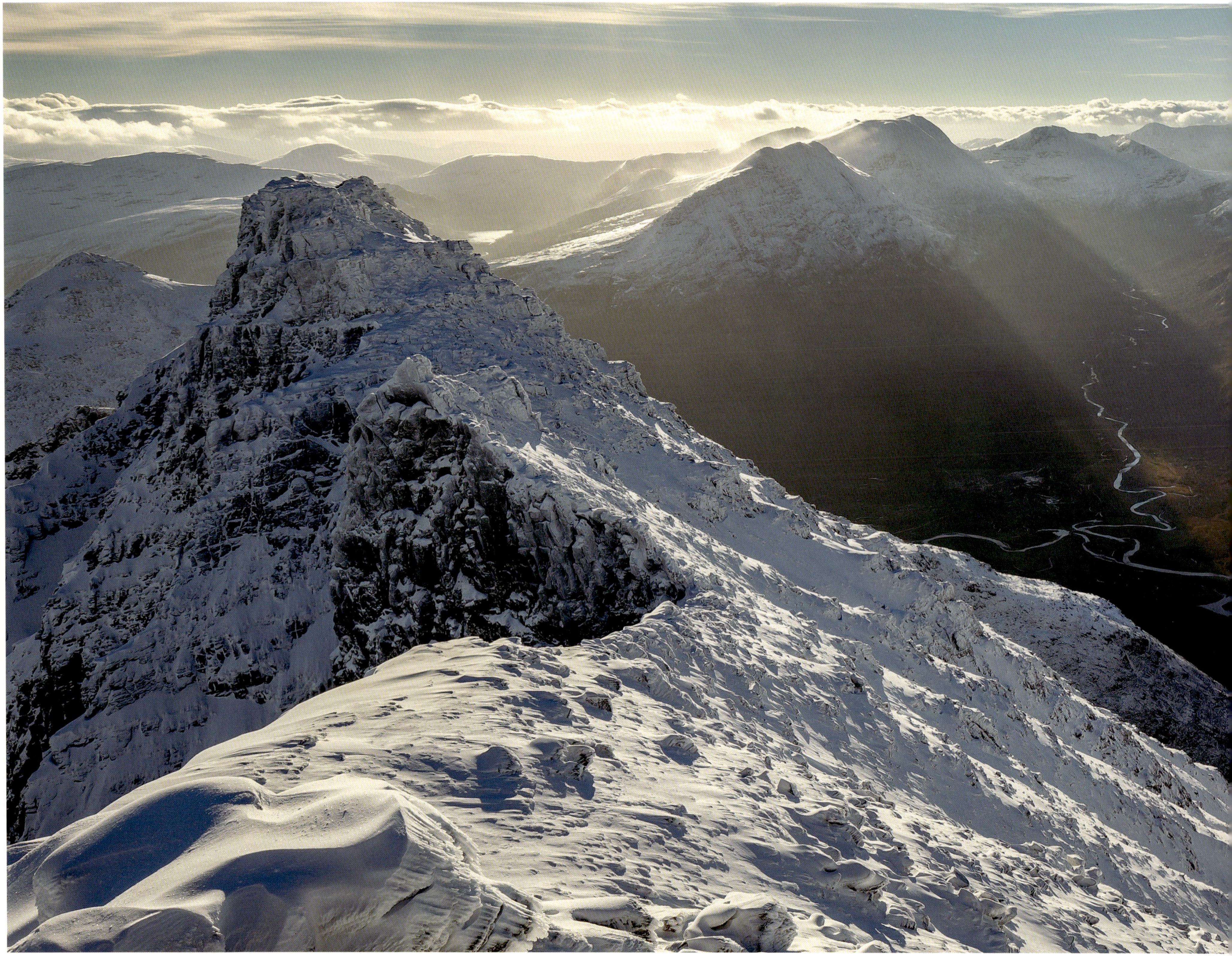

Loch Na Sealga and the
Fisherfields from Sgùrr Fiòna

A LESSON ON AN TEALLACH

Standing on the summit of Bidein a' Ghlas Thuill looking across to Sgùrr Fiòna I found it hard to focus on the task of taking photos. Before me was one of the greatest sights of my life, in the kind of conditions which landscape photographers dream of but rarely experience.

The serrated ridge opposite was steeped in snow, its aggressive faces sculpted by soft sidelight from the rising sun. In the distance I could see the familiar snow-capped peaks of Fisherfield and, at the edge of my vision, the mountains of Torridon. Passing gusts were blowing spindrift off the precipice, catching the sun's light as it fell. At my feet sat meringues of snow shimmering in the light. This moment was special. I looked down the slope to see Guy, my hiking companion on the trip, with a wide grin on his face.

An Teallach ranks amongst the finest mountains in Scotland. Compared to its neighbours in the Fisherfield Forest it stands as a true maverick. With an alpine character the massif is more impressive in both its scale and jagged nature than the surrounding peaks. The two most prominent ridges form a dramatic amphitheatre hinting at the mountain's name: An Teallach, 'The Forge'. The full ridge traverse takes you on a scramble along exposed crests and pinnacles making it one of the Highlands' more demanding hikes, especially in winter. But, given our relative inexperience and this being only my fifth day hiking in the snow, we opted for the safer 'there and back again' route from Dundonnell.

We started at noon on the previous day, ascending up a wet route which rises rapidly from the road. Before long the puddles beneath our feet turned to ice and patches of snow became a blanket of white which obscured the path altogether. We hoped to camp just below the summits to be within launching distance for sunrise and make the most of the short daylight hours.

Being the end of December the recent effects of Christmas had taken their toll on my fitness but we made steady progress to our camp spot at Sròn A' Choire. We settled in for the night while the summits remained resolutely in cloud.

Overnight the wind turned and strengthened, striking the tent from the side. The sound kept us up all night, our vulnerability all too apparent. We lay awake questioning our motivations and hoping that dawn would bring better weather.

When the alarm went off an hour before sunrise at 8am the final ascent seemed pointless. Usually I try to reach the summit thirty minutes or more before sunrise but with the peaks hidden in the cloud and still feeling incredibly tired we stayed in our sleeping bag cocoons.

Sometimes luck comes when you feel like you deserve it the most and with just twenty minutes to go until sunrise, the cloud began to lift off the mountaintops. Cue a panicked exit from the tent followed by a rushed hike up our first Munro of the day, Bidein a' Ghlas Thuill. The snow was firm and grippy, perfect for ice axe and crampons although admittedly we prioritised speed over technique. With sunrise soon upon us we paused just short of the summit to photograph overhanging cornices catching the earliest of the sun's rays (page 88). By 9.30am we were on top, enjoying scenes that will no doubt stay with us for a long time (page 89).

It took forty minutes to reach our final peak and in that time the wind had turned from a gentle breeze to a savage gale. The ridge looked incredible; snow was blasting horizontally off the edge in waves (below) but as captivating as the scene was, it felt like walking through a sand blaster. For the first and hopefully last time I had forgotten my ski goggles and as the weather worsened my ability to see deteriorated. We needed to get back. Guy and I had a difficult decision to make: either return the way we came and risk an exposed ridge walk or descend out of the wind down the incredibly steep slope below. We chose the latter. Mistake number two.

That afternoon, after descending Bidein a' Ghlas Thuill and heading up Sgùrr Fiòna, we found ourselves enjoying more towering views, this time over Loch Na Sealga, Beinn Dearg Mor and the Fisherfield Six (page 82).

Whilst we sat eating lunch, a hiker appeared a little further along the ridge on Lord Berkeley's Seat. He confidently picked his way along what looked to be a tricky route. Once he had reached us we learned that the winds were forecast to pick up soon, but we thought little of it. Rather than returning to the tent, I convinced Guy to extend the route just a little further to Sgùrr Creag an Eich. This was to be a telling mistake.

Initially we were well sheltered by the ridge but as we progressed downwards, the gusts increased. The slope became wind scoured and icy. We regularly stopped to brace as a blast of wind and snow hit us; all the time aware that any mistakes could have severe consequences.

It was a fearful descent, requiring a great deal of care and concentration. Twice I returned upslope to assist Guy, my more technical boots and crampons were much better suited to what had effectively turned into a low grade winter climb. Eventually, and with great relief, we arrived at the bottom of the slope and continued back to the tent.

There we found my 4-season mountain tent mangled by the wind, half buried, and shaking violently in the gale. The doors were torn open at both ends and the inside was full of snow.

We crawled desperately into the flattened shelter and used our backs to prop up the collapsed side, the wind fighting us all the while. I looked across at Guy, his face etched with worry - no doubt I looked the same. Moments later we burst into uncontrollable fits of laughter as our ridiculous situation finally dawned on us. Afterall, the worst was behind us.

As darkness fell we packed up, bundling the tent away as best we could and retraced our feint steps in the snow. At the pub in Ullapool that night we asked the barman if he had any local beers to which he replied: 'Yes, An Teallach'.

Cornice, Bidein a' Ghlas Thuill, An Teallach

Sgùrr Fiòna and the Corrag Bhuidhe Pinnacles from Bidein a' Ghlas Thuill

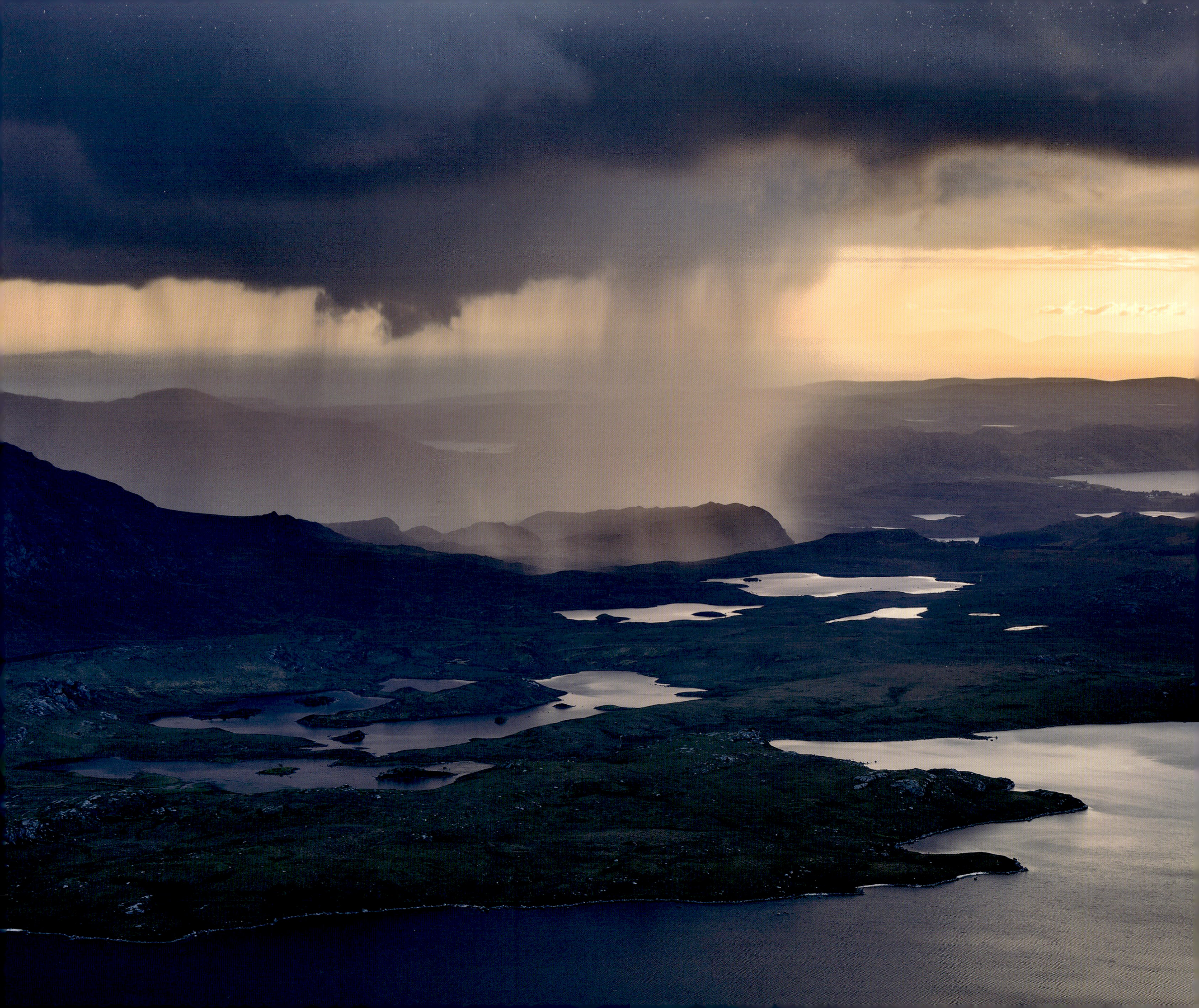

ABOVE: A' Mhaighdean from Beinn a' Chàisgein Mòr
OPPOSITE: Lochan Beannach Mor from Beinn a' Chàisgein Mòr

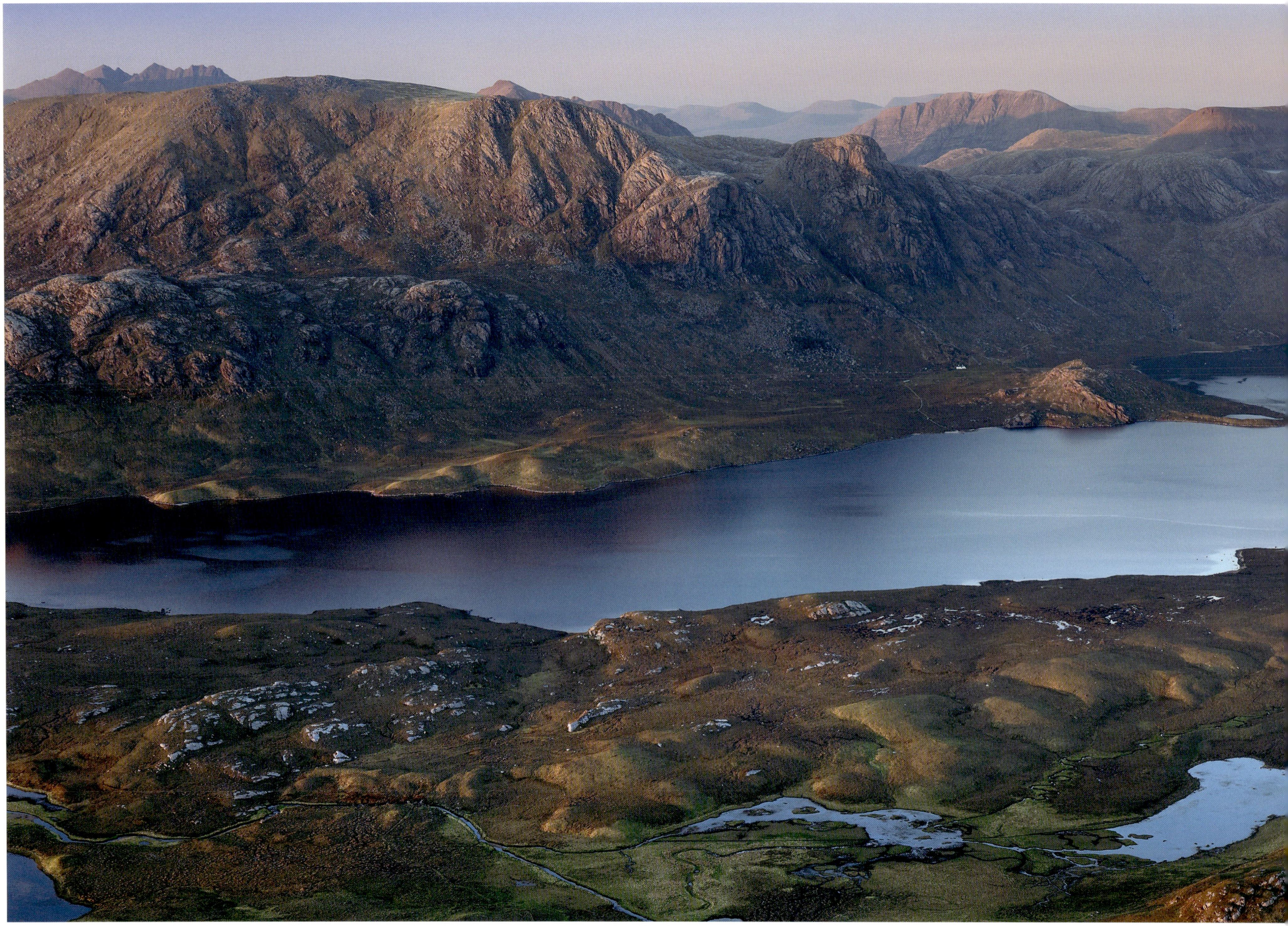

Fionn Loch and A' Mhaighdean from Beinn Airigh Charr

Gleann na Muice and An Teallach from Beinn Tarsuinn

Loch Na Sealga and An Teallach from Beinn a' Chlaidheimh

THE LOCH MAREE ISLANDS AND REWILDING

Loch Maree is widely regarded as the most beautiful loch in the Highlands. At 20km long and up to 4km wide it is a substantial body of freshwater, the fourth largest in Scotland. The slopes rising from the water are adorned with Scots Pine, notably in the Beinn Eighe National Nature Reserve. The loch is also surrounded by mountains, the most prominent of which is Slioch, The Spear, standing as a sentinel above the water, it's angular western face a favourite subject with photographers. But as scenic as Loch Maree is, it is on its wooded islands that magic is found.

A few years ago, on a perfect summer day, I packrafted out to the Loch Maree Islands with Emily. I went seeking unusual perspectives in the full knowledge that these pristine islands would undoubtedly have the potential I was looking for. We spent the afternoon paddling in our inflatable boats ocassionally stopping to discover a view or relax on a beach. In the evening as sunset approached the wind died down and reflections appeared on the water. I came ashore and began the midge dance - finding a composition, running away frantically swatting midges, returning, setting up my camera, running away, returning, taking a photo, running away... I have no doubt the image on page 101 completely fails to convey my distress, but not every emotion can be captured!

Whilst the experience that evening was memorable, it was the afternoon's exploration that affected me most.

Paddling between the Loch Maree Islands is like stepping back in time to an ancient Scotland. There you can find yourself surrounded by Scots Pine packed together so dense that it would be difficult to walk between them. On the fringes occasional dead trees stand like skeletons bleached white by the sun. Above the pines rise the peaks of Torridon and to the west the dominant rock face of Slioch. This is a landscape untouched by man, free from industry, deforestation and grazing, a wilder Scotland.

The charm of the Highlands has always been beyond doubt, its position as a world-class landscape firmly cemented in the minds of hillwalkers and tourists alike. But although wildness exists in the Highlands it cannot be considered truly wild.

In the mountains you can find solitude and beauty in abundance, but they are not untouched, nor is this an environment in a natural balance. The 'Fisherfield Forest' is its other title, yet there are few trees. This is a Deer Forest, a landscape managed for the purposes of hunting, a landscape grazed over centuries, an ecological desert.

As much as we celebrate the Highlands for its scenery, we must also recognise how deeply we have depleted nature. There has been a long history of deforestation in Scotland dating back to Neolithic times. Trees have been cut for fuel, timber and agriculture with forests shrinking in line with a growing population. Historically unprecedented numbers of red deer now ensure that the forests cannot return without human intervention.

Scotland's native woodlands now cover just 3% of their natural range, with the Beinn Eighe and Loch Maree Islands National Nature Reserve forming a tiny fragment of the once vast Caledonian Forest. These small woodlands represent a last haven for native species like red squirrel and pine marten as well as a wealth of plants, mosses and lichens. They are isolated paradises that allow us to reconnect with nature and give us a glimpse of the past and our possible future. But it is not enough to simply protect these areas, we need to expand and then reconnect these ecological islands and restore the natural wealth that once existed here in abundance.

In recent years rewilding efforts have gathered pace. Campaigns spearheaded by conservation and rewilding organisations have increased awareness of the multitude of benefits that come from restoring a landscape to its natural order. Encouragingly this message is also reaching policians and land owners.

High profile news stories about the possible return of apex predators like the lynx and wolf have created controversy drawing much needed attention to the cause. But rewilding is about far more than returning predators to the landscape.

Rewilding is about the large scale restoration of naturally functioning ecosystems allowing nature to do its own thing. It extends far beyond forests to marshland, blanket bogs, watercourses, wildflower meadows and high mountain heathland. It requires imagination, ambition and a step change in values amongst a public which is only now waking up to the importance of natural processes.

Rewilding presents a positive vision for the future. The benefits go far beyond increased biodiversity to provide solutions to a range of problems posed by the modern world. If you want to stop flooding downstream then look to beavers to build dams, or plant trees to absorb rainfall and delay water level rises. If you need a renewable fuel source plant more trees. If you want purer, cleaner water then send it through peatland or reed beds. If you want to reduce deer numbers, introduce their predators, the lynx and the wolf. If you want to absorb the carbon dioxide we're pumping into the atmosphere then plant more native forests and restore peatland. If you want a peaceful refuge to relax, to see abundant wildlife or to stimulate tourism then rewilding has the answers.

The Loch Maree Islands and Slioch

Garbh Eilean and Slioch

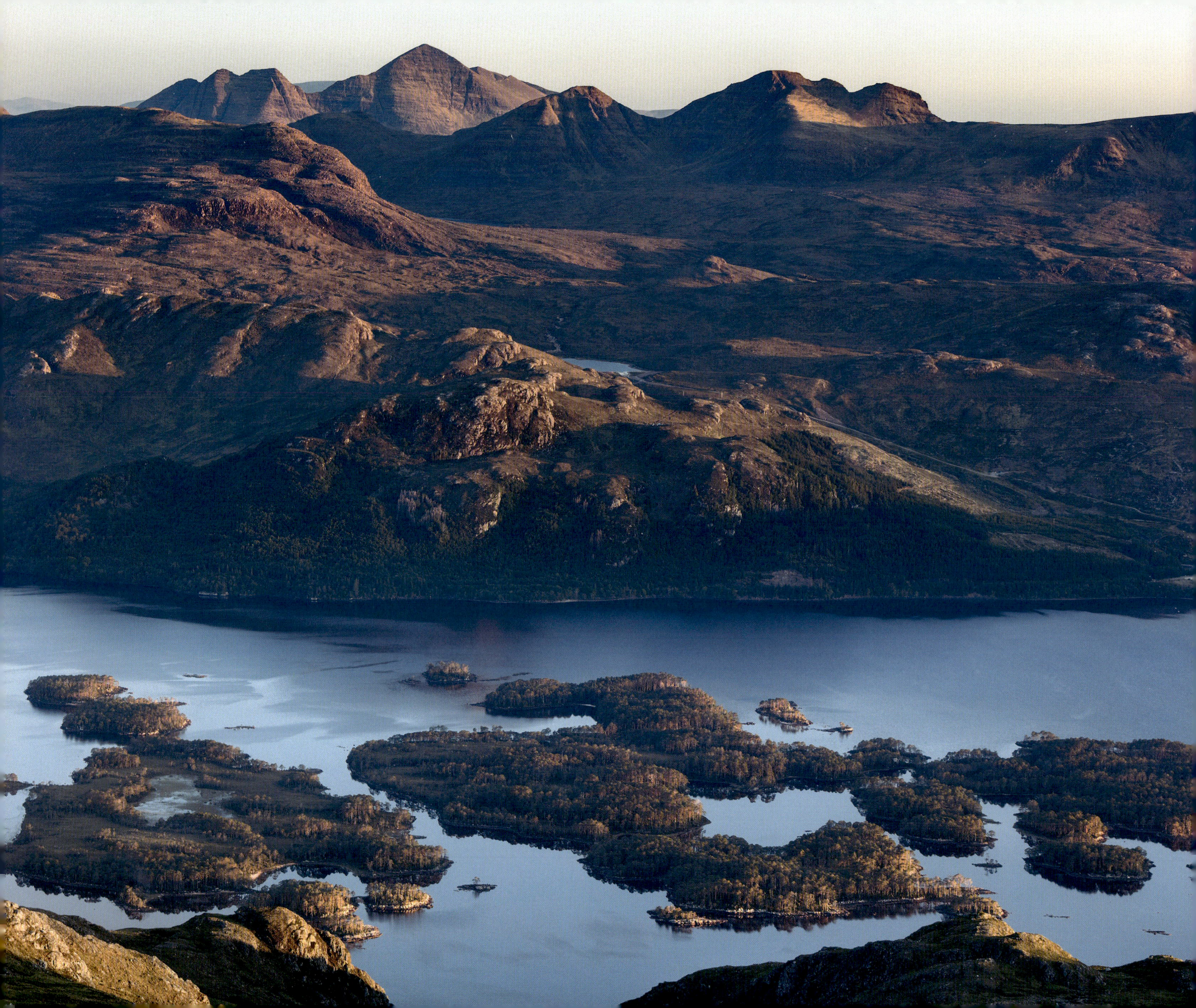

ABOVE: Tree stumps near Fionn Loch
OPPOSITE: Loch Maree Islands and Torridon from Beinn Airigh Charr

Buttresses of Beinn Làir from A' Mhaighdean

Allt a' Chladhain and Dubh Loch

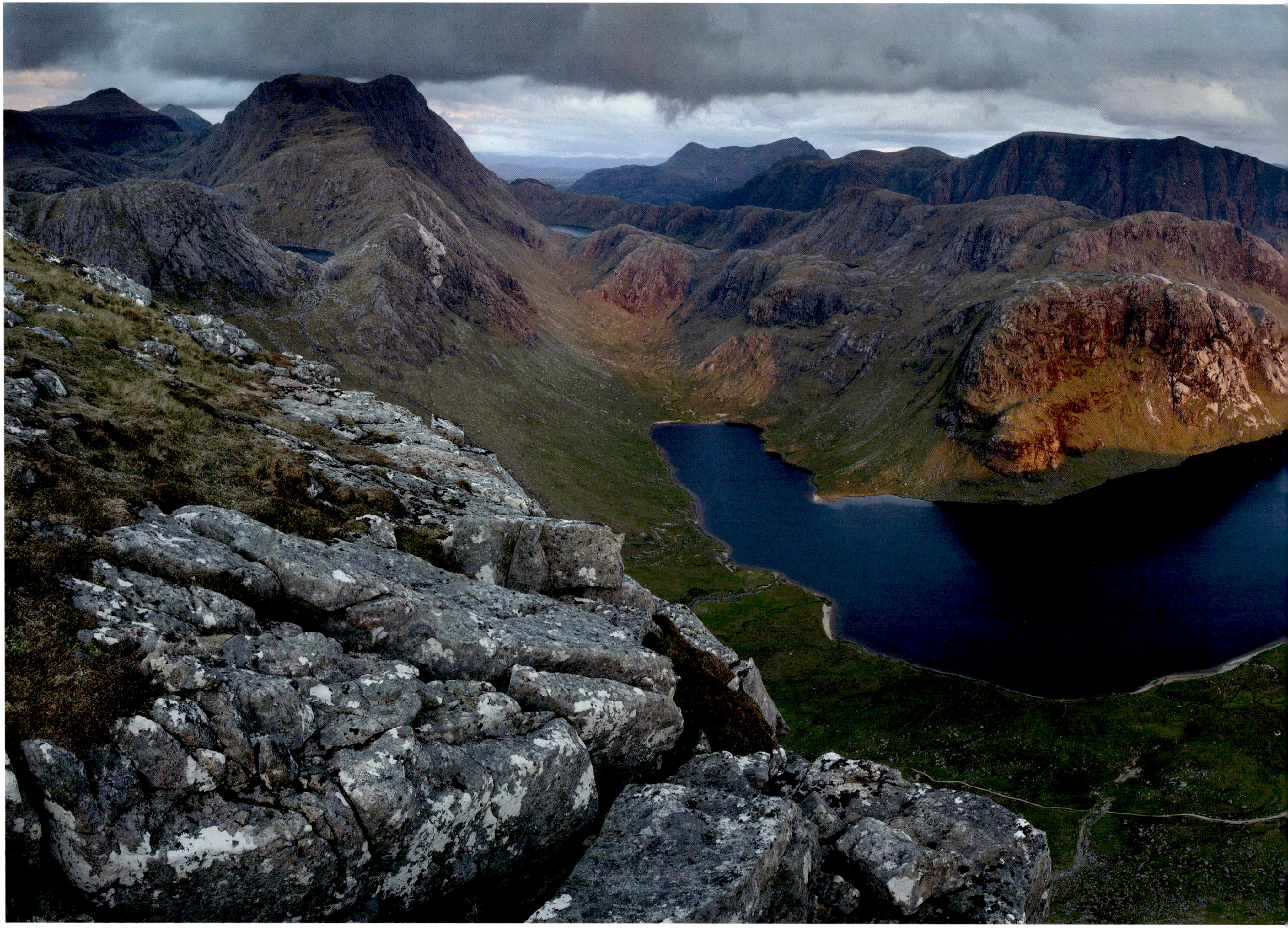

Ruadh Stac Mor, A' Mhaighdean, Slioch, Beinn Làir, Meall Mhèinnidh and Beinn Airigh Charr from Sgùrr na Laocainn

ASSYNT AND COIGACH

From Loch Broom to Kylesku

Assynt and Coigach are amongst the most unique of the Highland landscapes. Whilst the geological basis of the area is similar to the areas to the south - the mountains being largely comprised of Torridonian Sandstone - this landscape stands as an exception. The peaks are diminutive, rarely exceeding 650m of prominence from the surrounding ground level. Yet what the mountains lack in size they make up for in character and surroundings. Wide tracts of undulating ground separate the summits, with countless lochs and lochans interspersed between them. The mountains themselves are often precipitous, the southern buttresses of Cùl Mòr being particularly impressive. Stac Pollaidh is one of Scotland's easiest hikes and the most popular peak in the area perhaps due to its scarcely believable 360 degree views over a fantasy landscape below. To the north lies the area's crowning glory, Suilven, where a long walk is richly rewarded with inspiring scenes to the south over Fionn Loch and back to Stac Pollaidh.

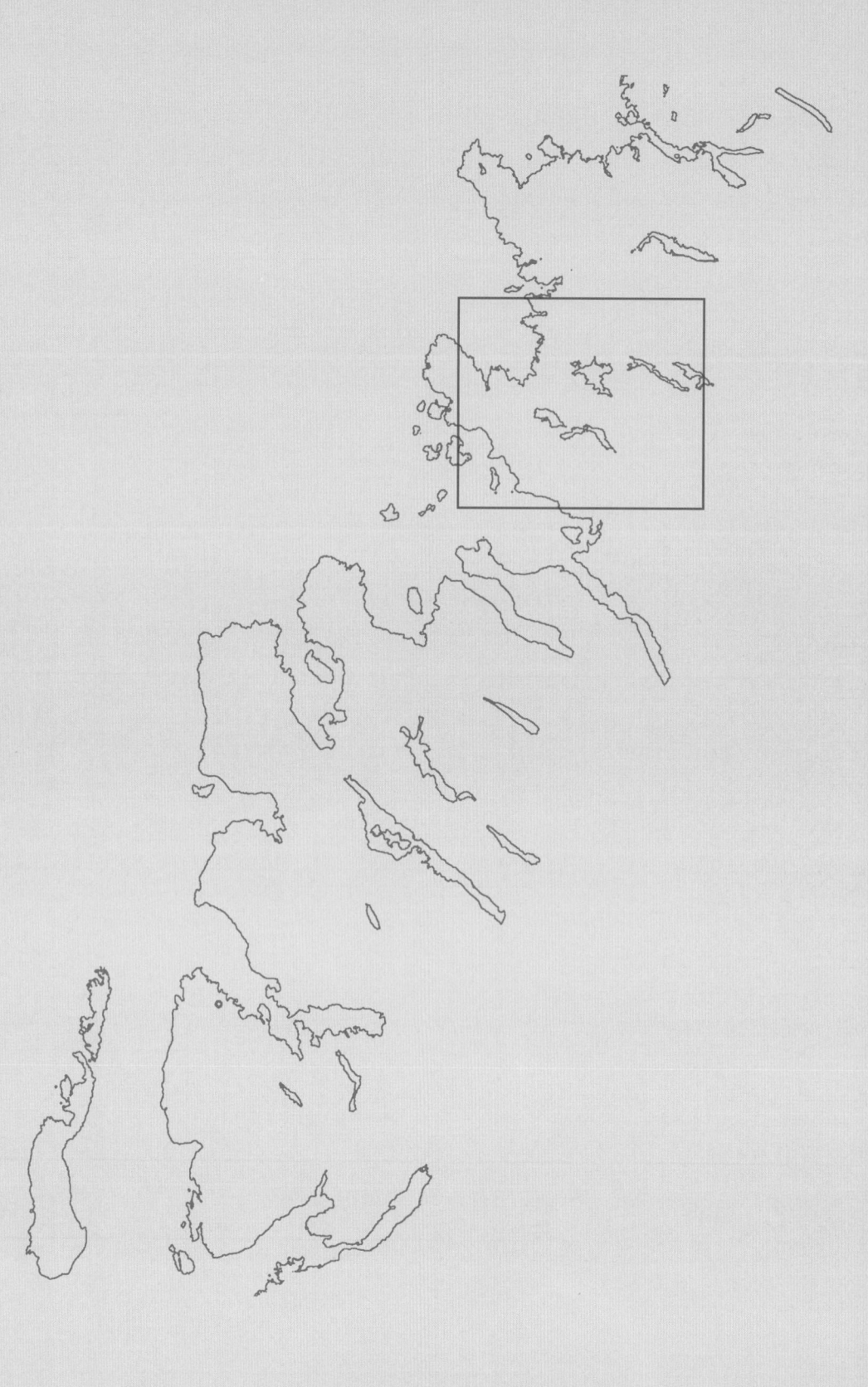

Map

- THE MINCH
- Lochinver
- Fionn Loch
- Suilven
- Canisp
- ENARD BAY
- ASSYNT
- Loch Sionasgaig
- Loch Veyatie
- Cam Loch
- Achnahaird Bay
- Loch Osgaig
- Cul Mor
- Stac Pollaidh
- Loch Bad-a'-Ghaill
- Loch Lurgainn
- Cul Beag
- Sgorr Tuath
- An t-Sail
- COIGACH
- Sgurr an Fhidhleir
- Ben More Coigach

FIRST TRIP TO ASSYNT

In 2009 I picked up a copy of Joe Cornish's book 'Scotland's Mountains'. Shot entirely on large format film it portrays the entirety of the Highlands in a series of thoughtful portfolios, each with it's own distinct flavour, accurately respresenting the variety of mountain landscapes in Scotland. It is, in my opinion, his finest collection of images and one of the most compelling accounts of the Highlands. It's a formidable book to have on the shelf now that almost a decade later I am publishing something similar myself!

Inspiration is a funny thing. As David Clapp once pointed out: inspiration rarely comes from your idols, but rather your peers. Your neighbour who gets out for every sunrise is far more likely to drive you to do the same than images of distant mountains from a photographer you have never met. Yet there is no doubt this book affected me.

One image was particularly striking; a summer view from Suilven in Assynt looking south over endless lochans towards Stac Pollaidh. The grassy hills were a verdant midsummer green and the lochans below shone like opals. At the top of the frame a group of mountains gave the scene an immense sense of scale. It was a fantasy world, a place I could scarcely believe existed, and it was in Scotland. My imagination jumped a mile.

My first visit to Assynt, and to the Northwest Highlands as a whole, centred around seeing and photographing this iconic view for myself. In the months prior to the trip I poured over Joe's images, enjoying them for their creativity, variety and spectacle. I studied maps of the area, plotted routes and virtually flew around computer generated mountains to the point that I could recognise and name every peak.

Finally one April I set off with my friend Jake, a fellow photographer, on the long drive north. Suilven was to be my first Scottish summit but Jake had never camped in the wild before, and certainly not on a mountain. I was an overconfident 23 year-old with minimal hiking experience and Jake was a concerned novice.

It was a long walk in to the base of Suilven for us uninitiated hill walkers and the final portion of the approach was boggy and energy sapping (although fortunately a path has since been built). This was exercise enough for Jake and he was considering camping at the bottom - I was certainly tempted!

The ascent itself was steep and loose, the unfamiliar weight of our rucksacks contributing considerably to our struggle. When we reached the bealach between the twin peaks of Suilven we still had a final climb to gain the true summit. I was glad for a rest by the time I reached the top but Jake was broken. When I hear the word 'tired' I sometimes think of Jake in a starfish position on his back on the grassy dome of Suilven that April. He stayed there for some time.

The scene to the south was, without question, the greatest view I had ever seen. Amazingly it exceeded the expectations of Joe's photograph. I realised for the first time that capturing the true grandeur of the mountains might forever remain an unreachable goal, but one worth aiming for.

Overnight the wind was wild. My cheap tent struggled to remain standing exposed as it was on the summit of the mountain. After a few hours lying awake with the tent buckling around me I stepped out to reset the guylines. A near freezing gale made short work of the little warmth remaining in my clothing, the combination of fear and cold causing me to shake violently. Once back inside I continued my sleeplessness resolving to buy a new tent and warmer clothing - a valuable learning experience.

I resurfaced just before dawn to a deep blue twilight. Patches of low cloud drifted over the rolling hills below. Suilven's second peak, Meall Meadhonach, took on a dramatic pyramidal shape from our viewpoint and above it the crescent moon drifted slowly southwards. It wasn't the scene I had come to photograph but the resulting image is one of my favourites to this day (page 123). It was a seminal moment for me as a photographer, the point at which effort and reward became an irresistible combination, when summits became addictive. The point at which I found a new way forward.

My photographs of the view to the south fell short. A noticeable haze posed problems throughout our time on the mountain. The sky was a pale grey and the rich oranges of the winter grasses below were washed out and lifeless. As Jake and I walked back to the car underneath sunny skies I was already plotting a return.

I tried again in the following years, sometimes cancelling trips due to bad weather and once ascending only to leave empty handed. On one notable trip I took a workshop group to the base of the Suilven, but on arrival the mountain was enveloped in cloud. There was little reason to camp on the summit so instead we hiked around to the north-west to find an unusual perspective. That night we experienced perfect twilight conditions with cloud drifting over the peak of Caisteal Liath (right).

Despite the setbacks, I hadn't given up on the view. Finally one May evening I felt I had captured an image that could close the loop, a suitable homage to an image that had brought me so much enjoyment (next page). It's one of few occasions I've specifically visited a location with the intent to capture my own version of someone else's photograph, and certainly the only time I have taken such a dogged approach, but it brought with it an immense sense of satisfaction.

On that occasion I was much better equipped, far more experienced and with a group in tow - a marked development from the first visit. We hiked in sunshine throughout and reached the top in plenty of time. An evening that started with few low clouds developed into a full cloud inversion just before sunset (right) and after the light had left us we sat on the grassy summit sharing a bottle of wine between the group and wondering if the Highlands could ever be better.

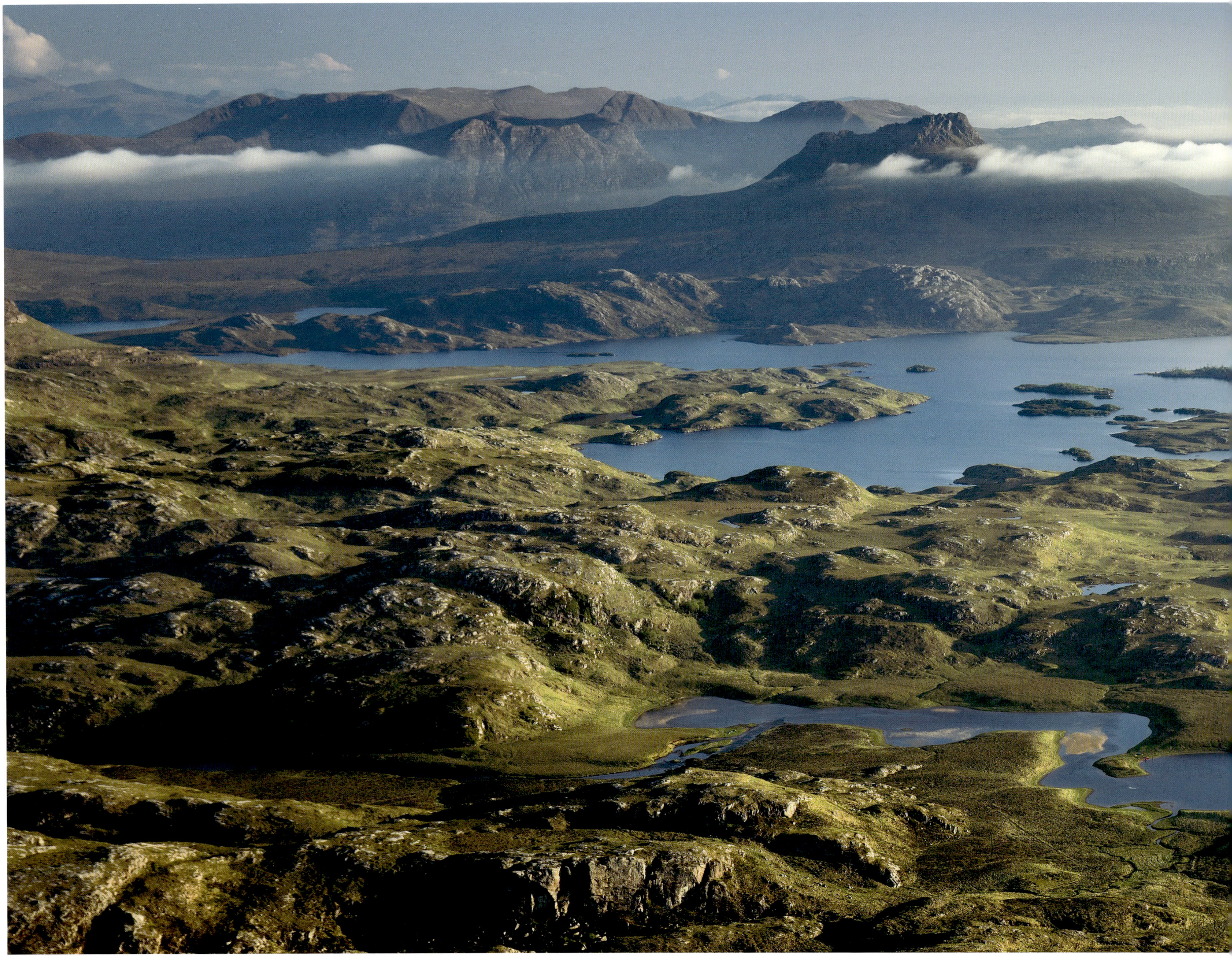

Fionn Loch, Loch Sionasgaig
and Stac Pollaidh from Suilven

Quinag from Loch na h-Innse Fraoich

The hills of Assynt and Coigach from An t-Sàil

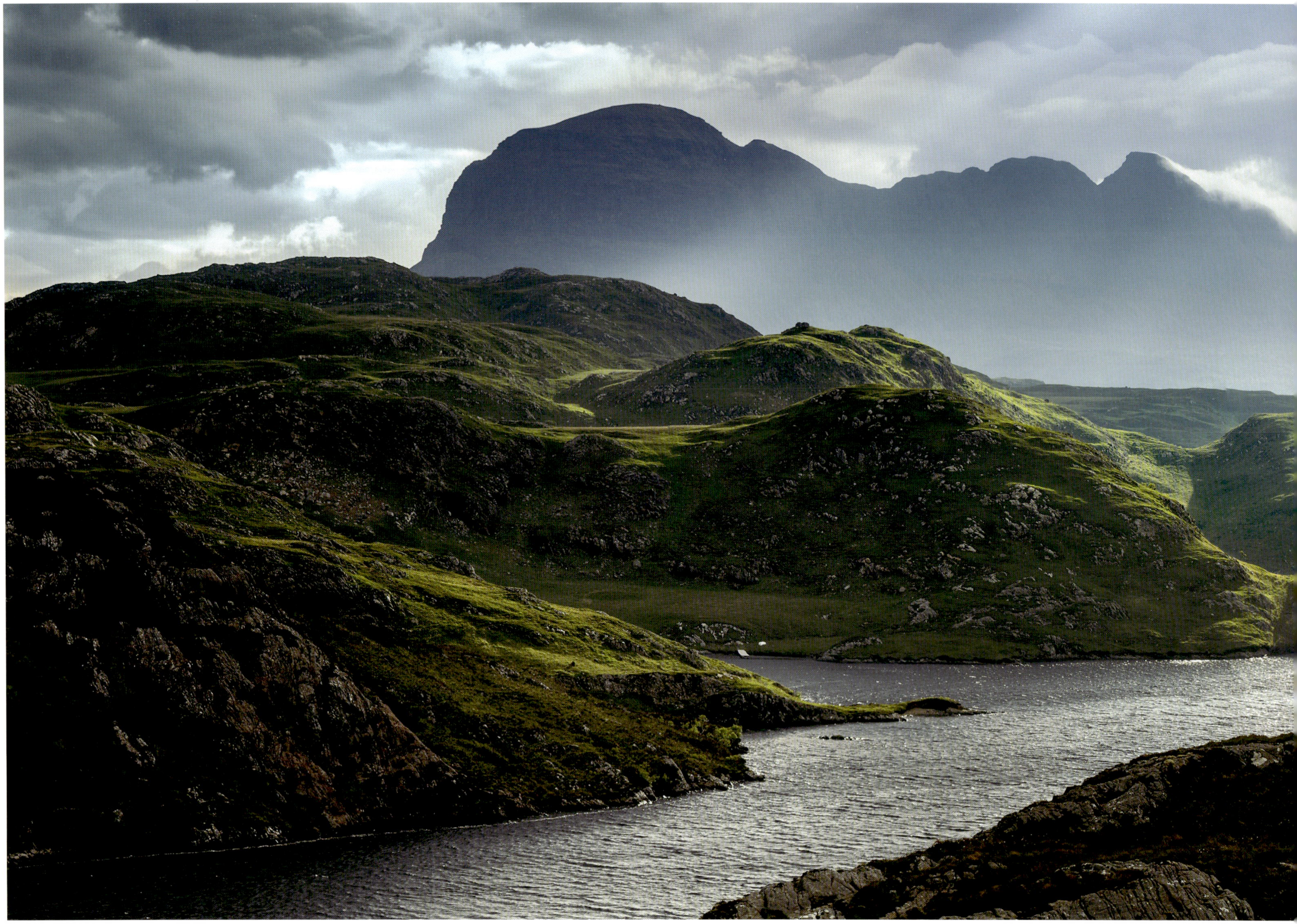

Suilven and Boat Bay

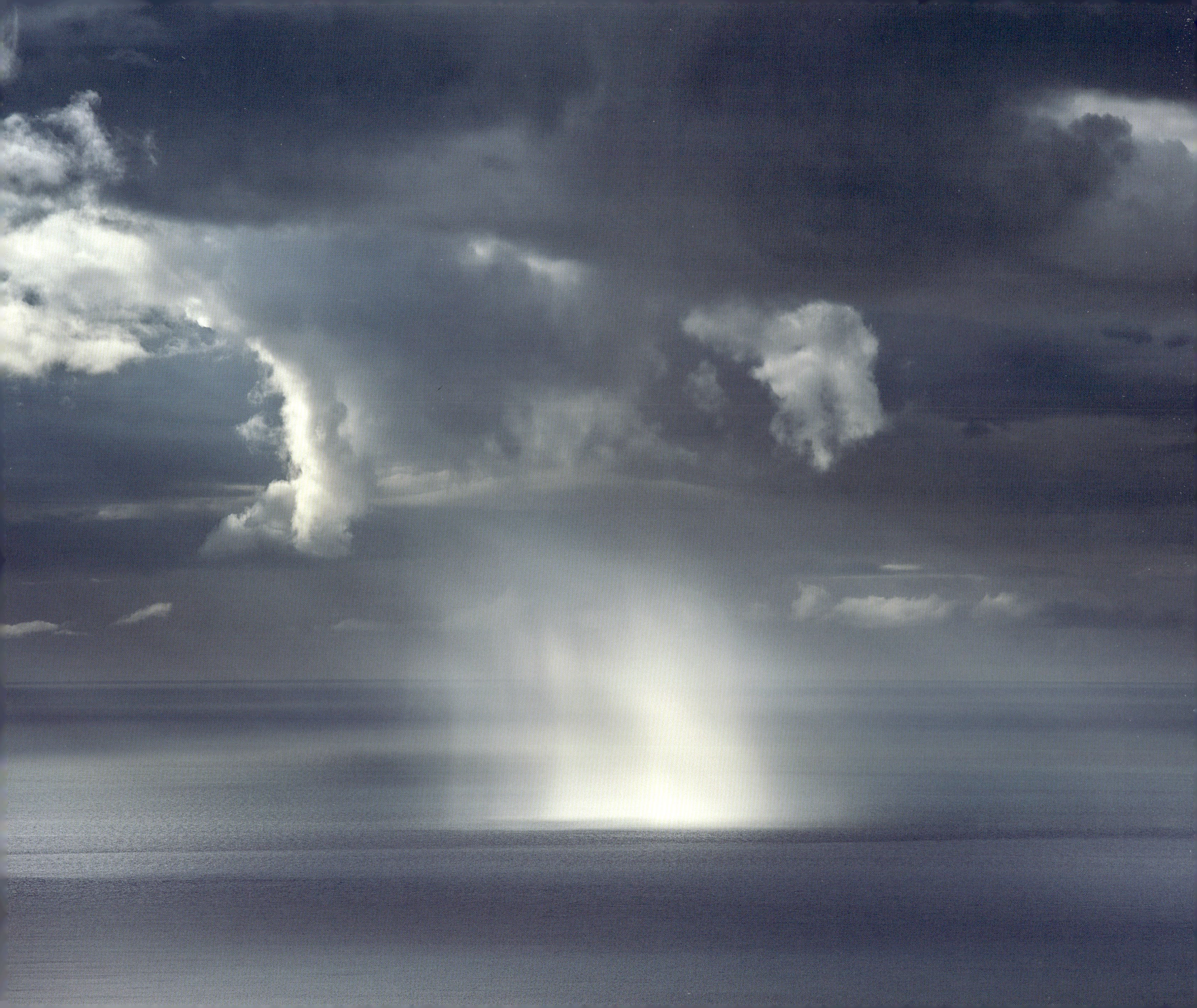

ABOVE: The Summer Isles from Ben Mor Coigach
OPPOSITE: Shower over The Minch

Achnahaird Bay and the hills of Coigach

Meall Meadhonach from Caisteal Liath, Suilven

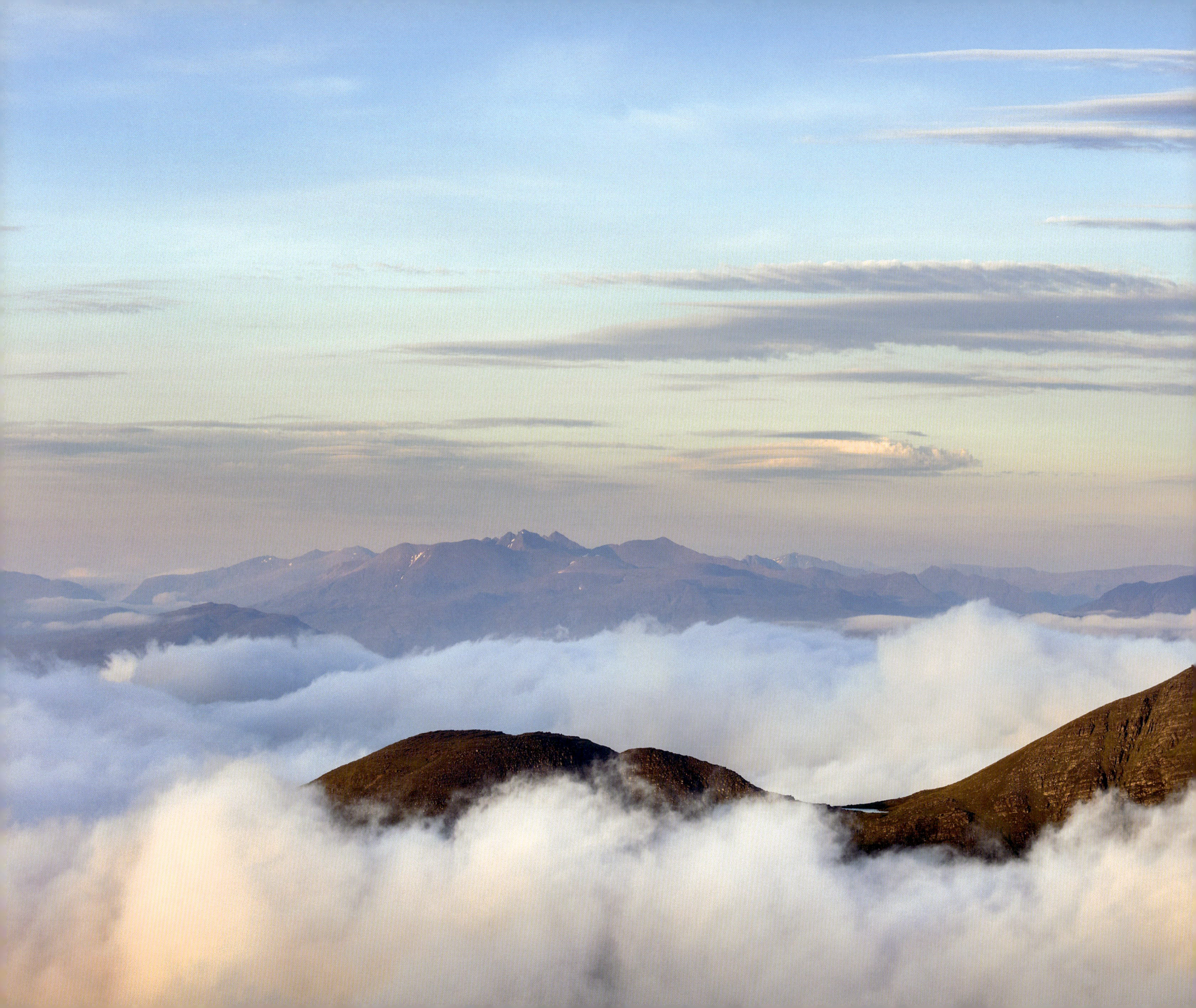

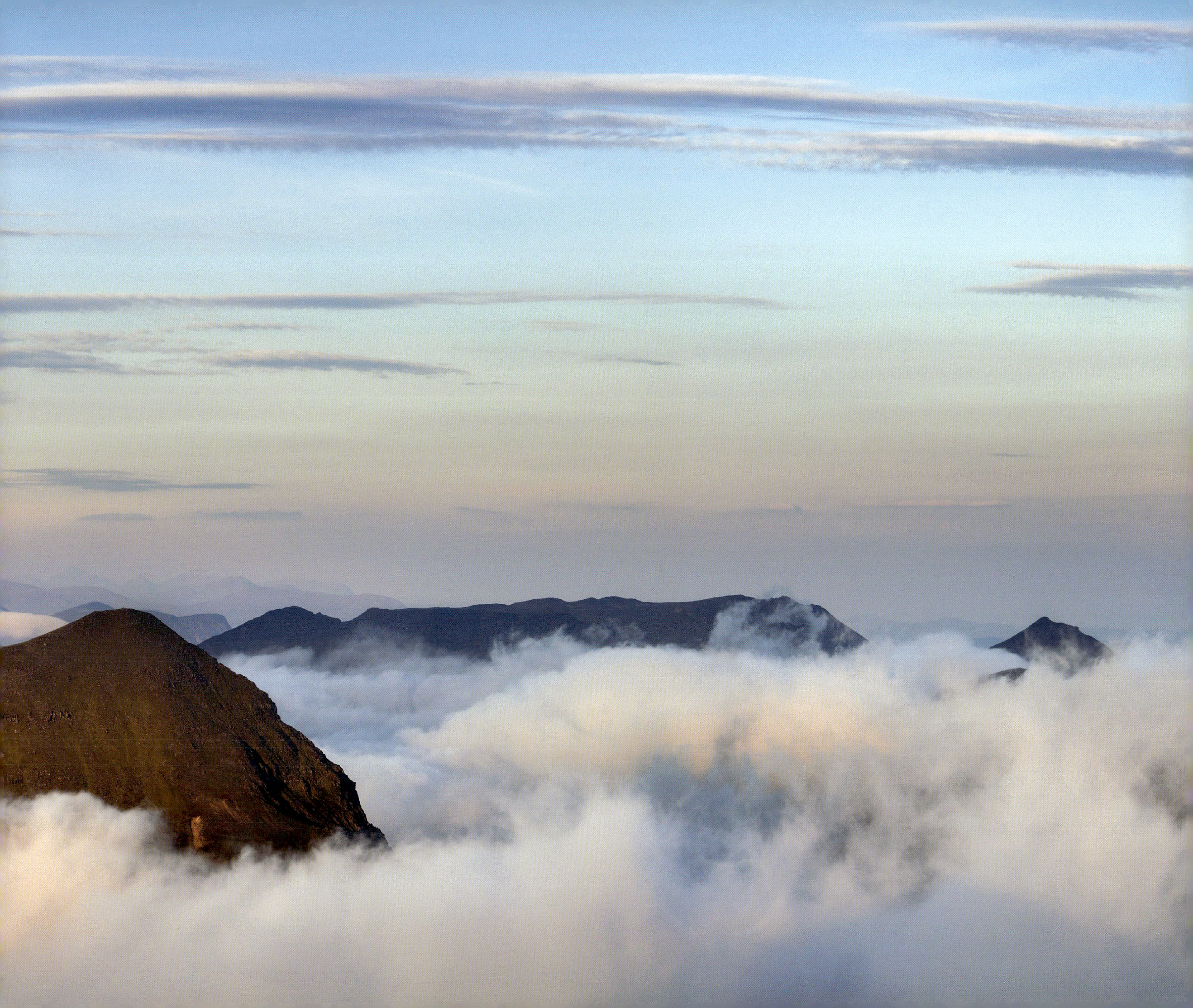

ABOVE: Torridonian Sandstone, Cùl Mòr
OPPOSITE: Stac Pollaidh from Cùl Mòr
PREVIOUS: Cùl Beag, Beinn Mor Coigach and An Teallach

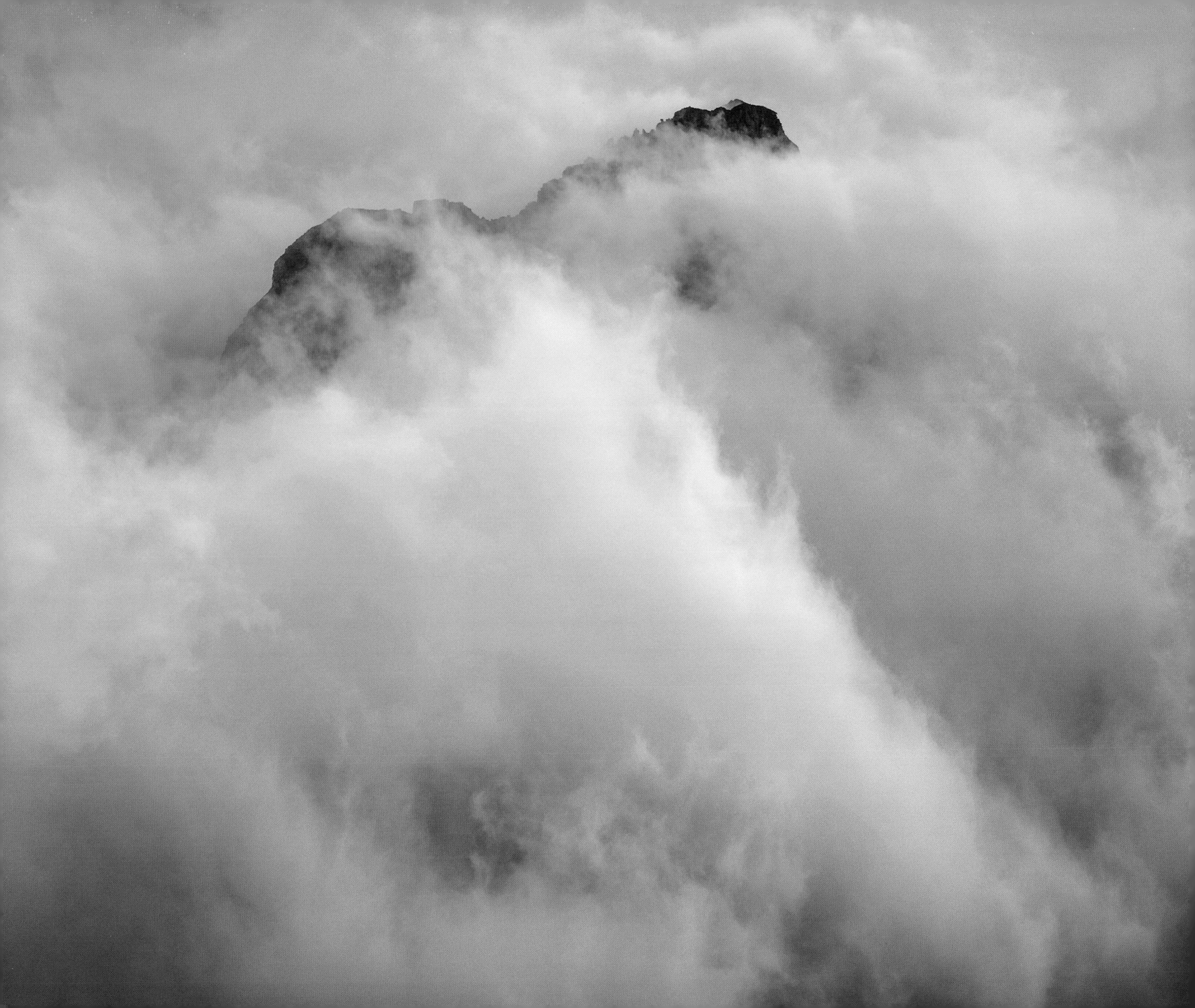

Beinn Tarsuinn and Lochan Tuath
from Sgùrr An Fhidhleir

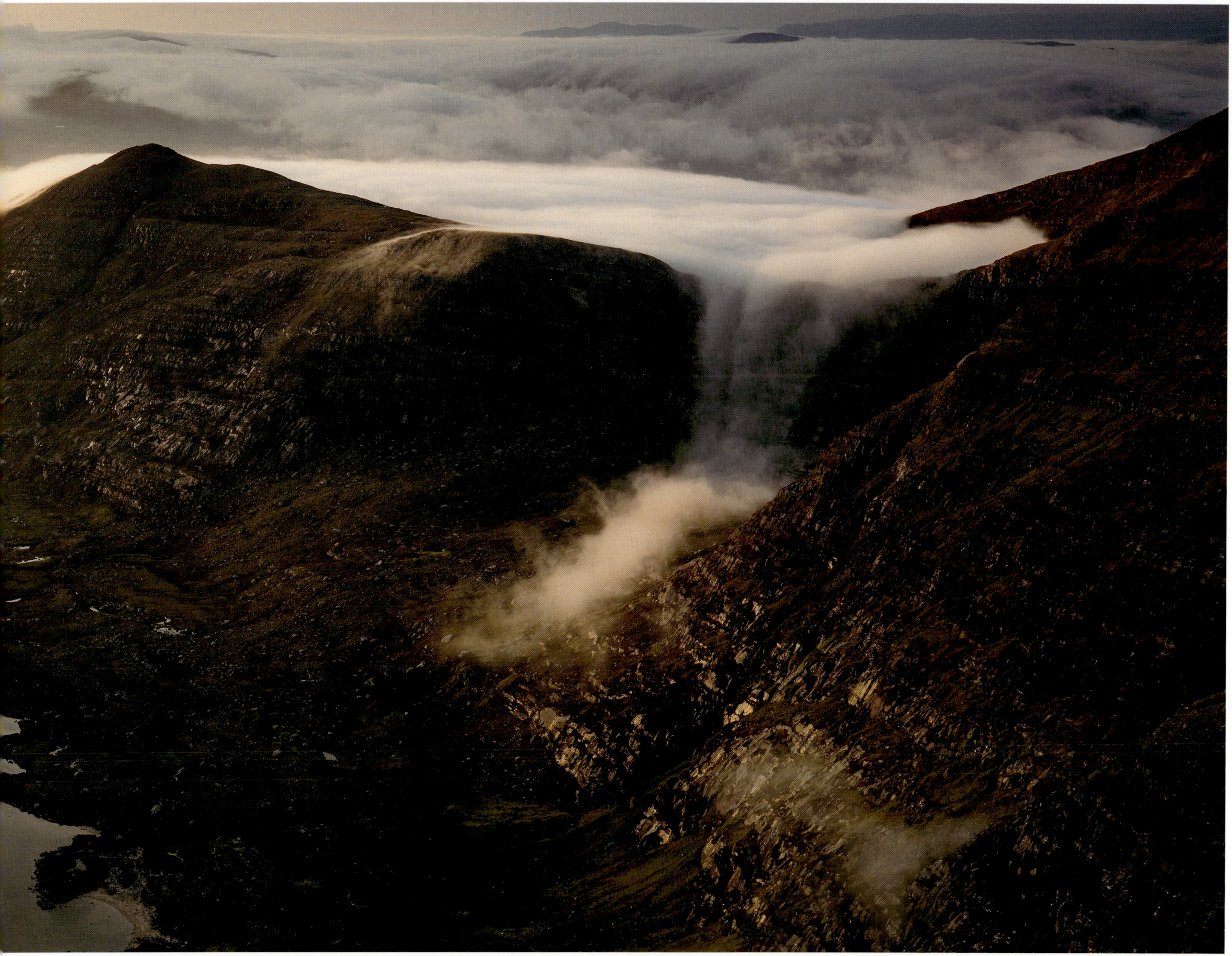

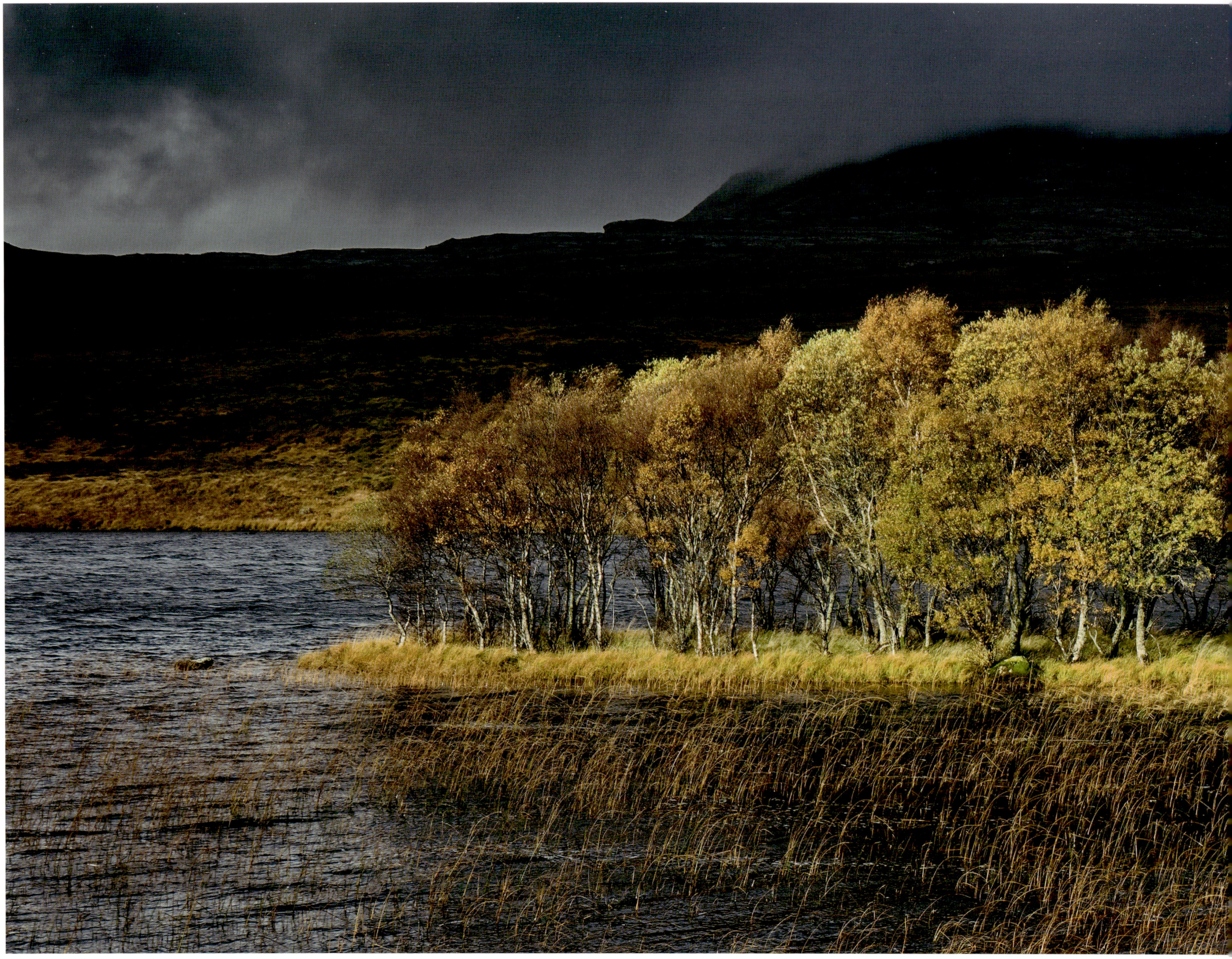

Birch trees, Loch Awe

Falls of Measach in heavy rain

Loch Buine Moire, Cùl Mòr and Cùl Beag

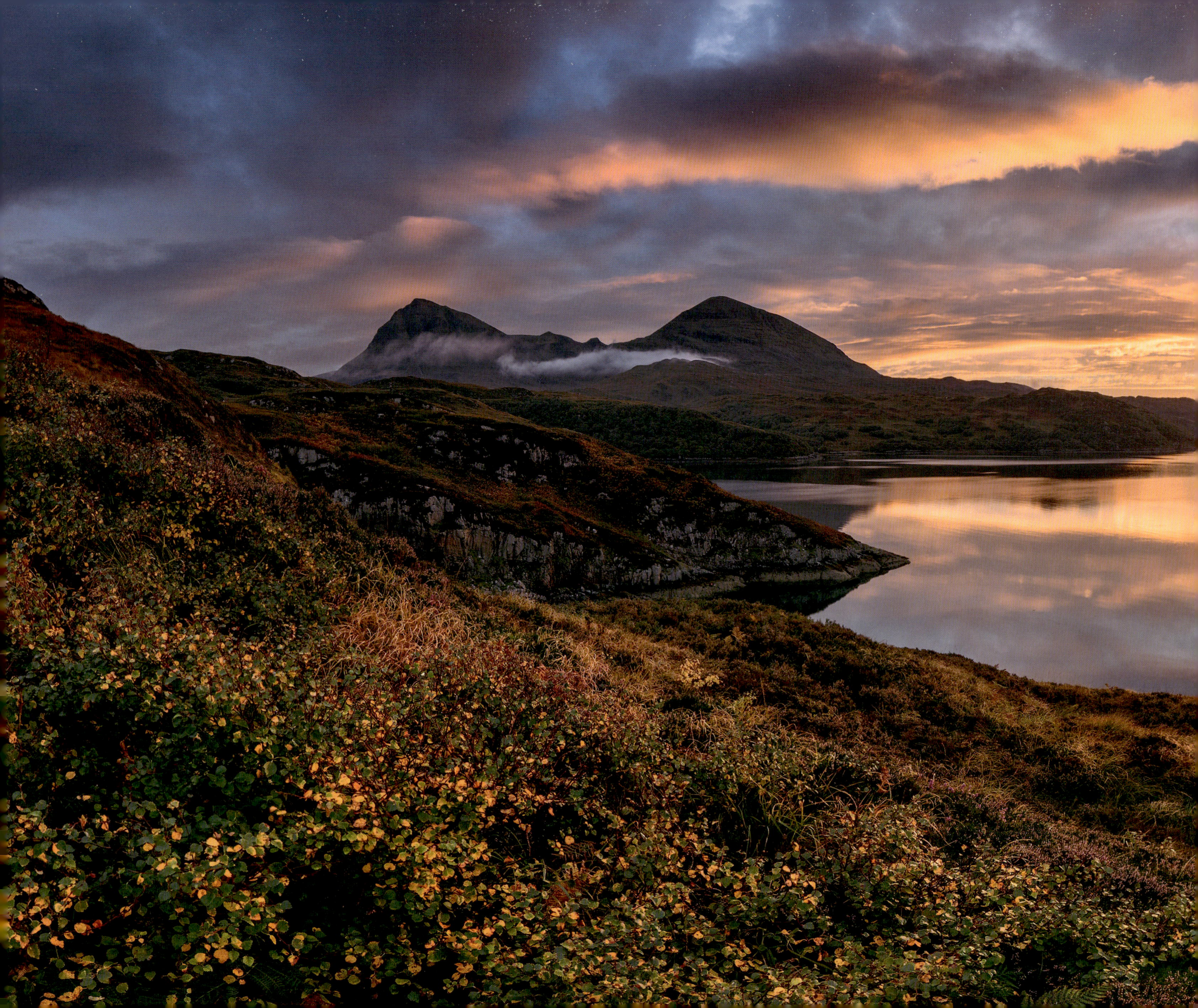

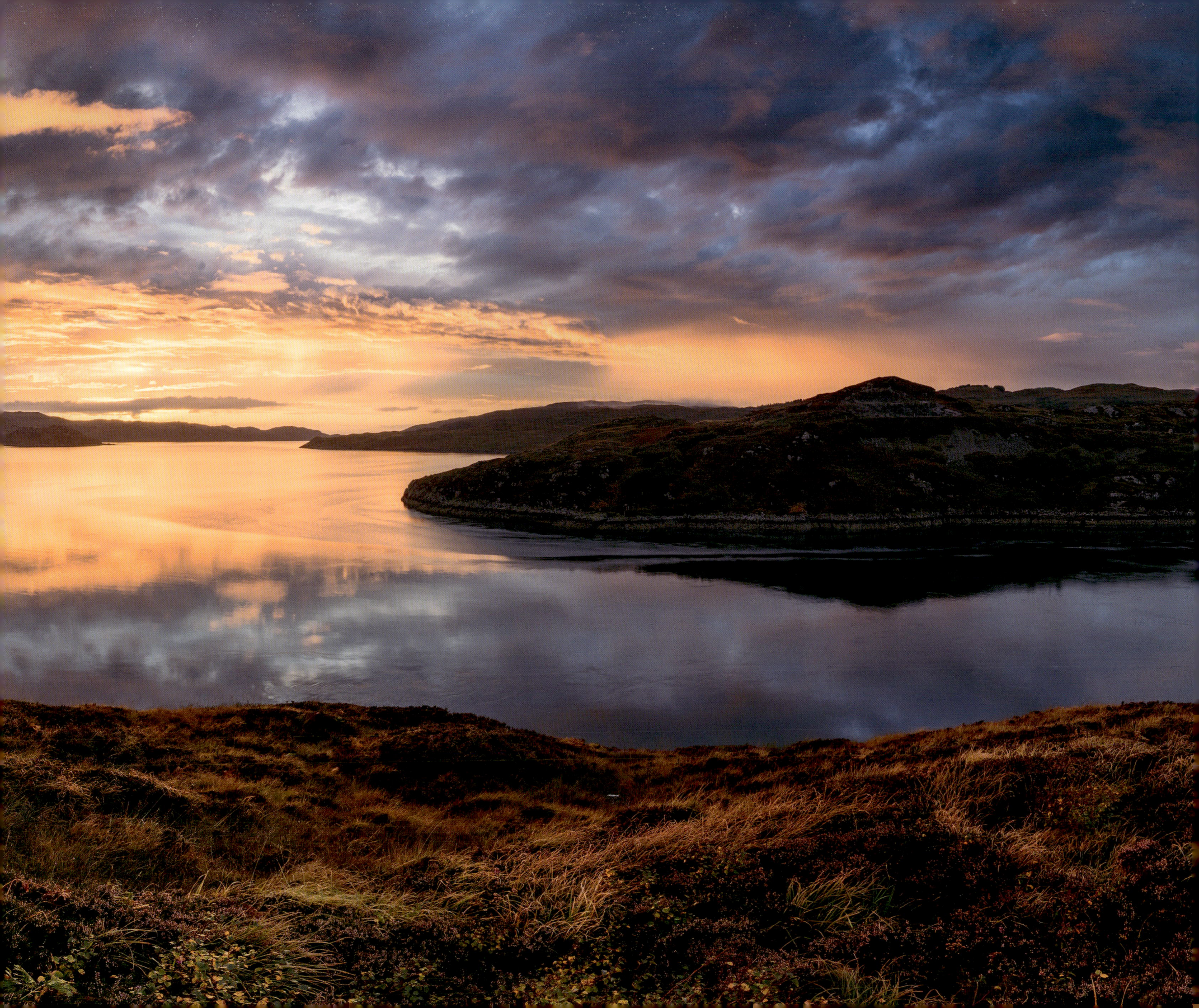

LEFT: Suilven from the west
OPPOSITE: Suilven and the River Kirkaig
PREVIOUS: Quinag and Loch a' Chràirn Bhàin

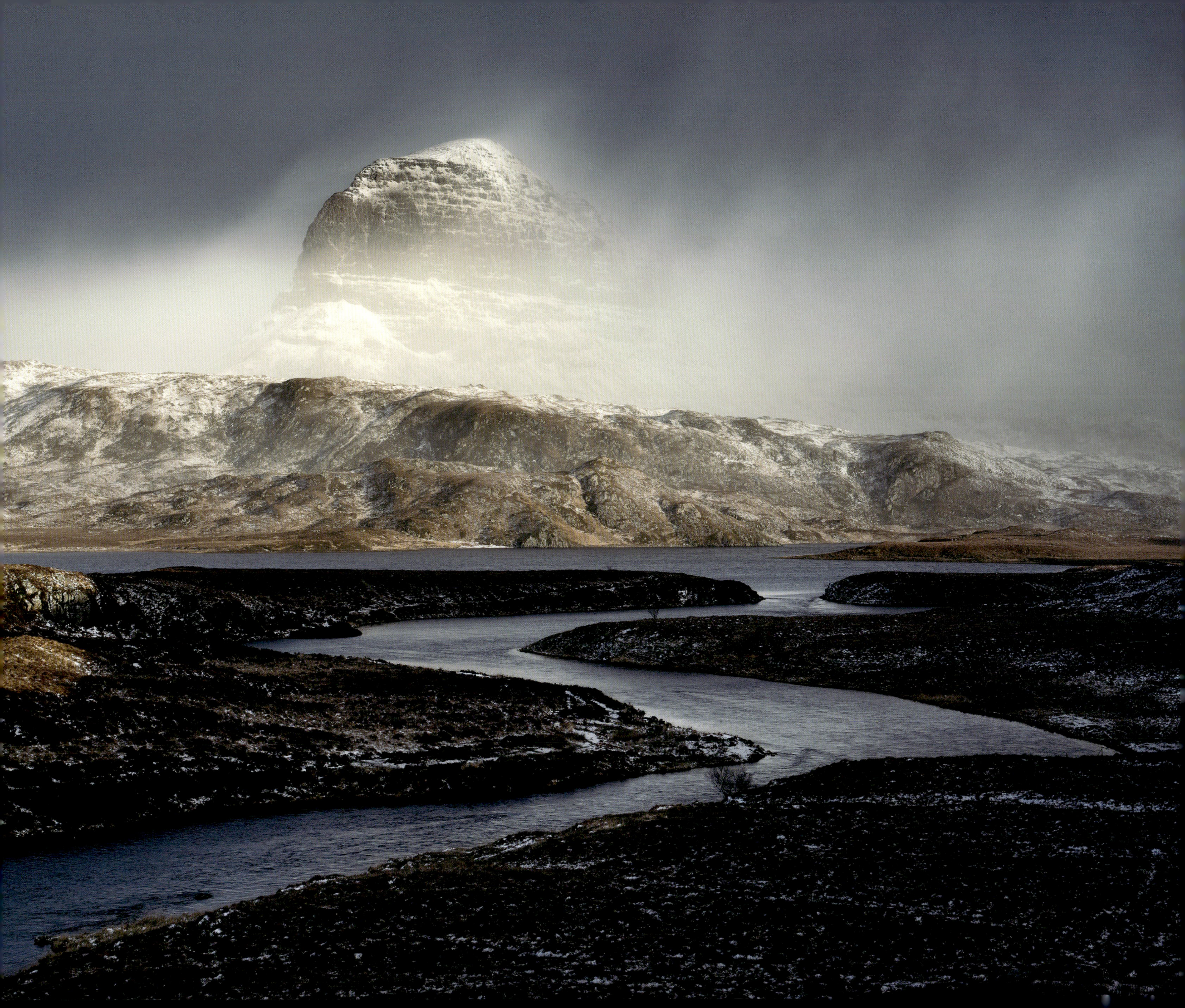

Glacial erratic and the hills of Assynt and Coigach

Sgùrr An Fhidleir across Loch Bad a' Ghaill

Sàil Gharbh, Quinag

141

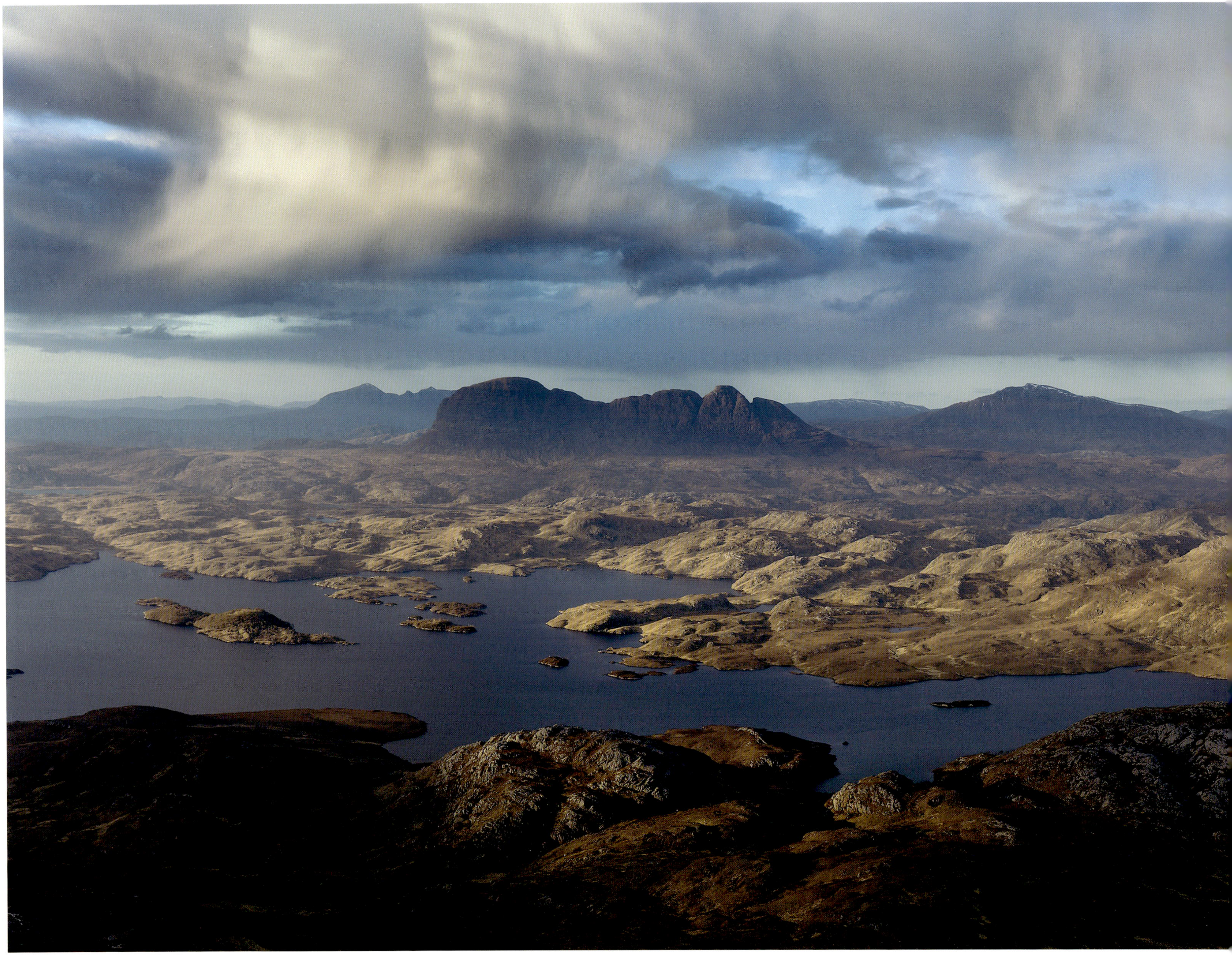

Suilven and Cùl Mòr
from Stac Pollaidh

ABOVE: Cùl Beag across Loch Sionasgaig

OPPOSITE: Stac Pollaidh from Sgòrr Tuath

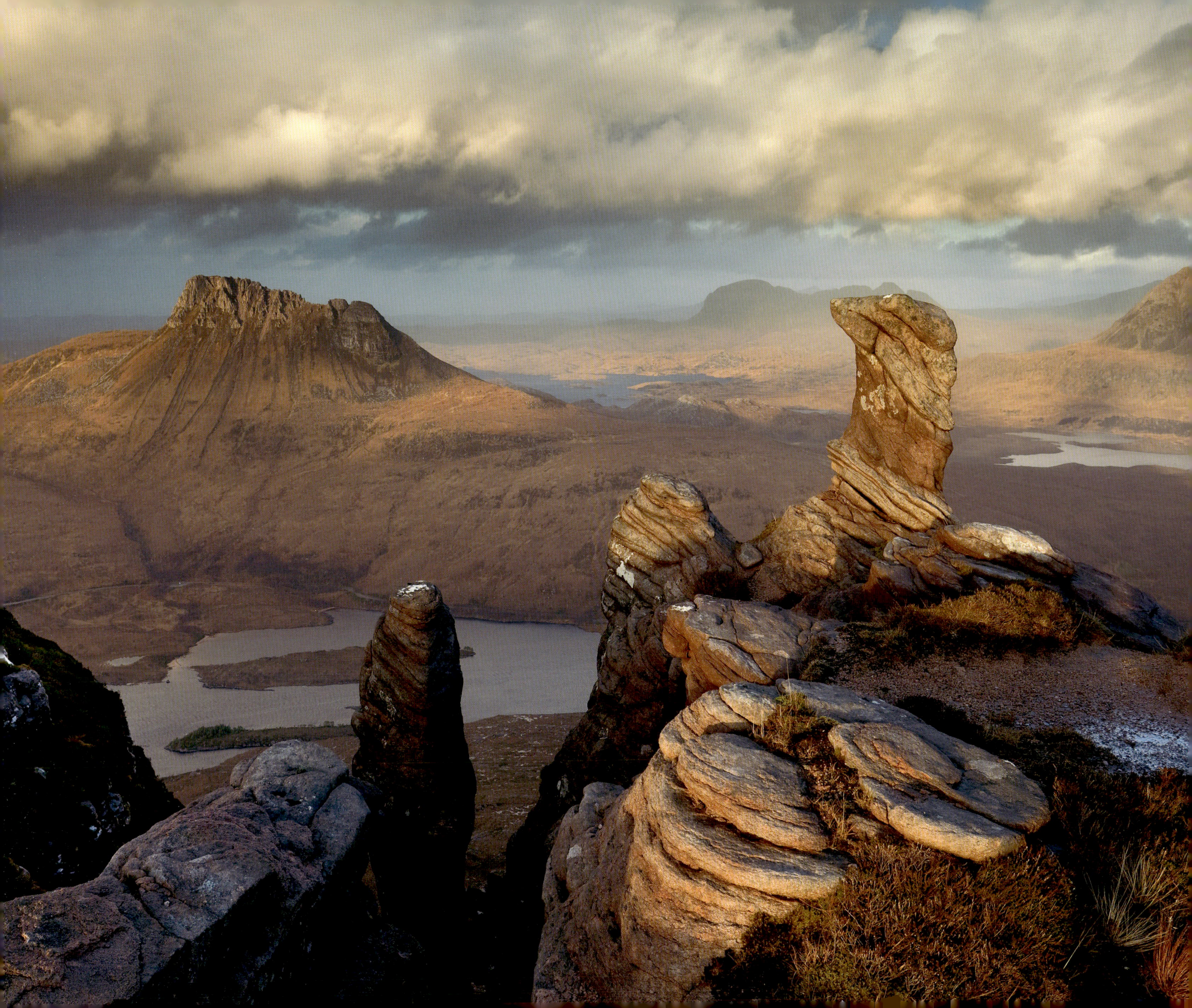

ACKNOWLEDGEMENTS

The greatest thanks must go to my girlfriend Emily, my parents Claire and Vaughan, my twin brother Lloyd, and my wonderful Nanna Margaret. You provided confidence when I doubted myself and unfailing support over the years, not least when rashly pursuing my dream as a photographer. I love you all dearly.

To my hiking companions Harsharn Gill, Guy Richardson, Hamish Frost, Sven Soell, Geoffrey Ball, Steve Sellman and Jake Spain, thank you for the memorable adventures and laughs along the way. To Harsharn, Hamish and Sven, thank you for keeping your images a secret so that I could present a book of largely unseen work. For that I am very grateful.

To my workshop participants too numerous to name specifically, thank you for your willingness to visit new locations and for serving as a regular reminder of the child-like wonder the Highlands can provide.

To Chris Townsend thank you for flattering me by agreeing to write the foreword. I hope you will continue to inspire hill walkers and the general public alike.

To Tim Parkin, Colin Bell and John McMillan thank you for your expertise and experience, you made this book a practical reality. To Tom White thank you for finding so many mistakes in my writing! To Jono Renton, thank you for your outstanding design work, feedback and keen eye for detail.

147

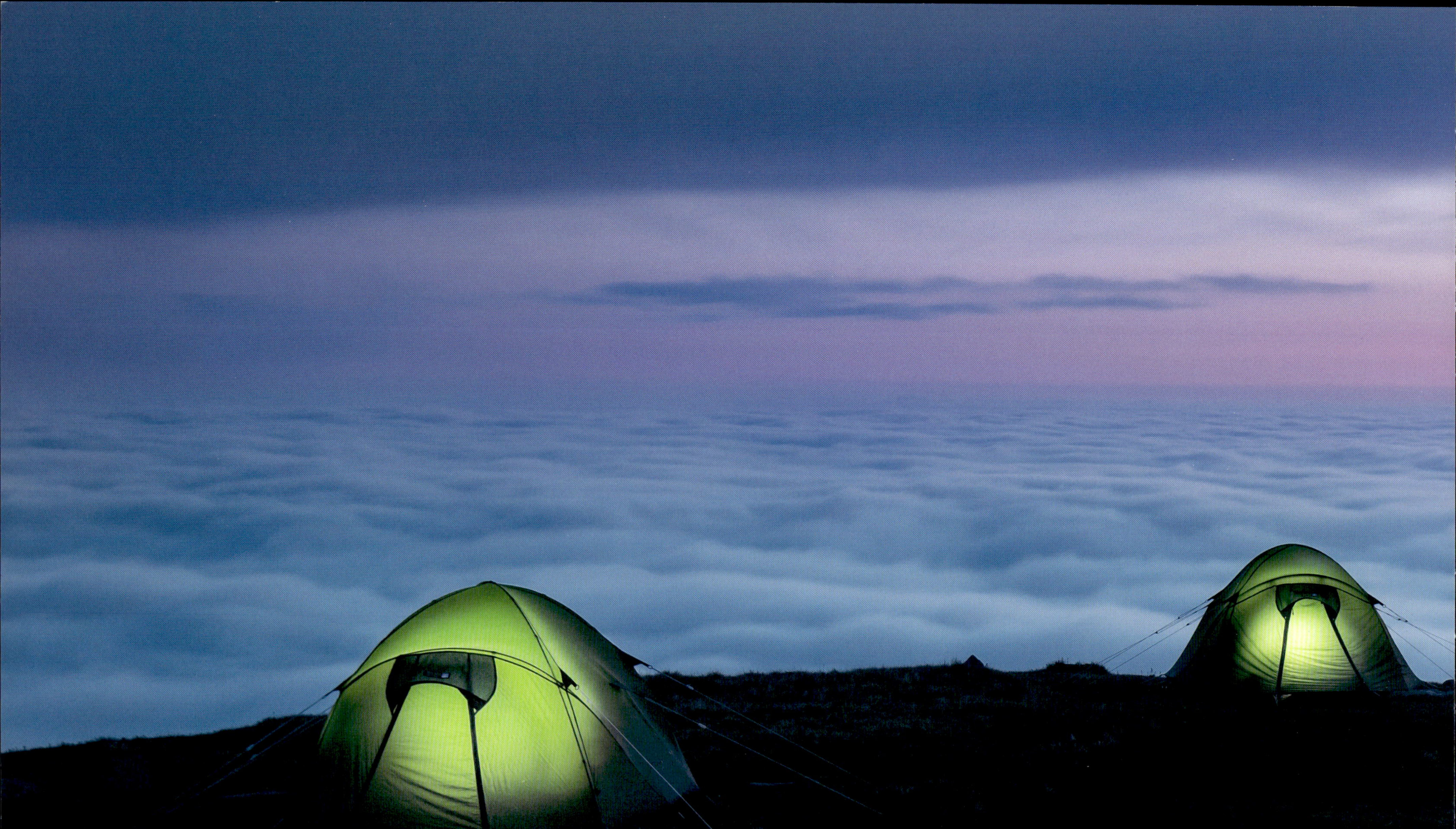

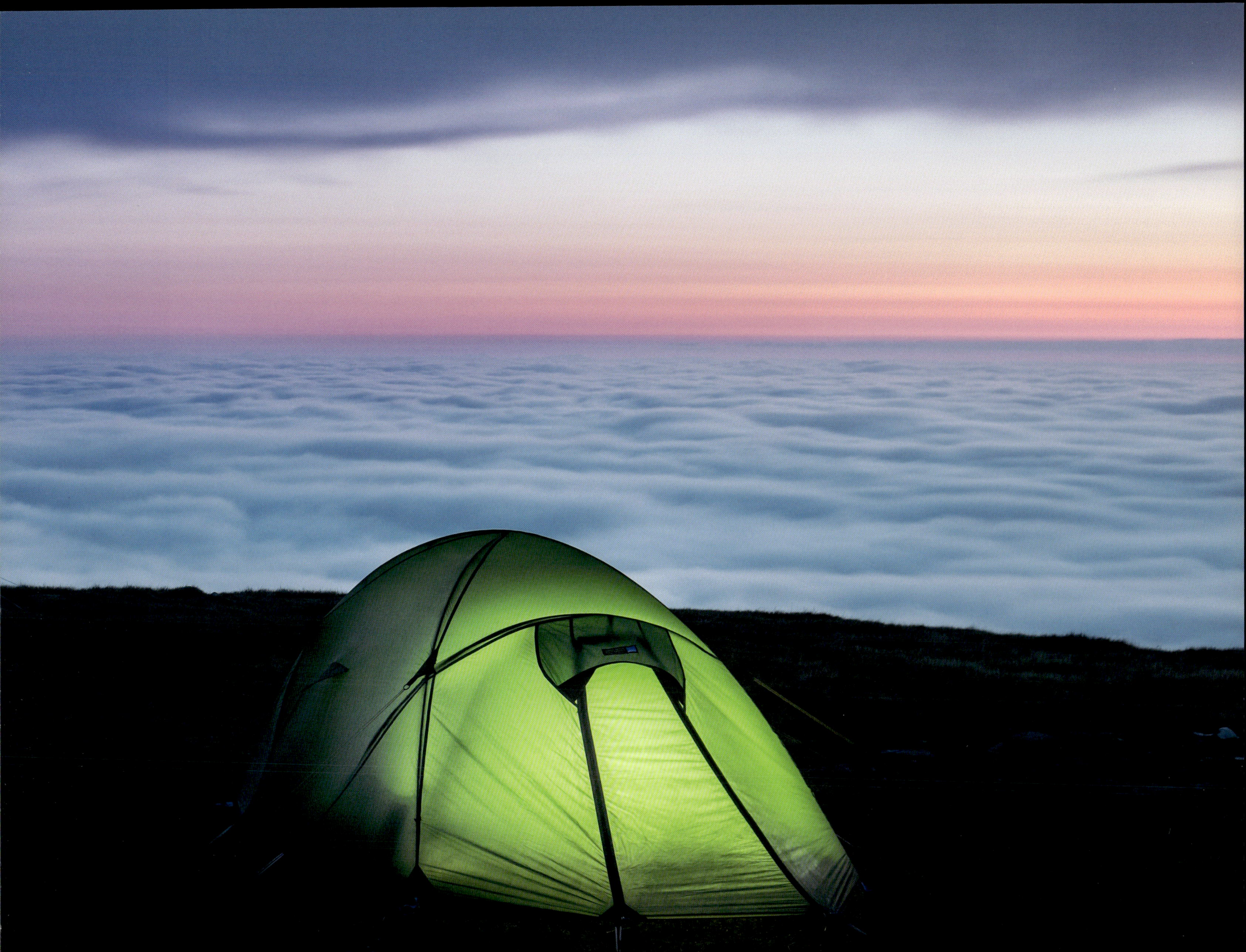

My sincere thanks to all those who pre-ordered this book. You helped to make it a reality.

Sean Freeman
Jan Pusdrowski
Colin Bell
Nick McLaren
Paul Zizka
Andrew Yu
Matthew Wood
John Altringham
Paul Taylor
Brian Lackey
Will Milner
Alex & Catheryn
Nicolas Alexander Otto
Nick Livesey
Paul Arthur
Jamie Gillies
Mark Phelan
Julian Cartwright
Omar Jabr
David Battensby
Joe Rainbow
Tim Allott
Alan Coles
Chap Lovejoy
Steve McGuire
Jonathan Philp
Sven Saelens
Alan & Hazel Wood
Brian & Margaret Turnock

Craig & Susan Adams
Danny Wootton
Ryan Simpson
Jörg Frauenhoffer
Kenny Muir
Maciej Markiewicz
Arild Heitmann
Rob Oliver
Tim Nicholls
Stewart Hunt
Jake Turner
Bobby Lee
Geoffrey Ball
Dan Melville
Jesús Roncero
Ryan Szydlowski
Mike Prince
Jean Londergan
Paul & Khris
Nigel Morton
Colin Daniel
Ollie Pocock
John Ash
John Bonney
Jonathan Carroll
Andrew Morehouse
Philipp Lutz
Julianne M Hartzell
Ross Brown
Richard Lizzimore
Ross Saxby
Don & Dawn
Mark Quinn
Jacomina Wakeford

Shaun Young
Milos Lach
Niall Henderson
Dani Lefrançois
Paul Gaughan
Mark Voce
Doug Urquhart
Robert Cronk
Daniel Groves
John McSporran
Tim Nevell
Al Miles
Howard Klein
Marc Hermans
Carla Regler
Jayne Sanderson
Murray Wilkie
Carl Brindley
Justin Nugent
Michael Prince
Cameron MacKay
Mark Mc Mullan
James Davies
Richi Abad
Seamus Crawford
Heather Tonge
Andrew Leaney
Steve Byrne
Tony McEwan
Graham Lawson
David Marshall
Jon Whiteley
Stefan Blawath
Stewart Smith

Graeme & Judith
Alex & Susannah Mason
Tommy Leggate
Michael Murphy
Tomas Frydrych
Teàrlach Coull
Sandra Bartocha
John McCarthy
Dave Mead
Paul Wayman
Gill Hood
Daniel Martin
Reto Steffen
Dave Fieldhouse
Michael Houghton
David O'Brien
Bruce Little
John Beatty
Adam Perfect
Philip Stewart
Grant Hyatt
Alex & Rosie
Glen Moyer
Thomas Happe
Xing Kai Loy
Jonathan Grundy
David Hecker
Isobel Henley
Lesley Bradley
Warwick Lloyd
Kersten Howard
Joe White
Jan Zwilling
Chris Grieve

Antoine Auchabie
Paul Marcellini
Tony Simpkins
Phil Sayer
Andrew Wilson
Norman Smith
Michael Sykes
Christopher Smart
Michael Vandenesch
Steven Hanna
Mark Griffin
Nicole Heise
Alfonso Salgueiro
David Tolcher
Sandris Grivins
Daniel Secrieru
Beat Meier
John McGuckian
Jeremy Rowley
Jose Gray
Andrew Briggs
Kev Keenan
Daniel Long
Graham Dunn
Katherine Saunders
James Bell
David Eberlin
Sophie Carr
Jan Moerings
Stéphane Beilliard
Tarn Stroud
Craig Marchington
Brian Doyle
Ron Coscorrosa

Tim Wrate
Katherine Saunders
James Bell
David Eberlin
Sophie Carr
Jan Moerings
Stéphane Beilliard
Tarn Stroud
Craig Marchington
Brian Doyle
Ron Coscorrosa
Tim Wrate
Iain Young
Simon Cunningham
Matthew Cattell
Erin Babnik
Euan Ross
Simon Owens
Piaras Kelly
Peter Edwards
Ben Ivory
Michael & Sue
Greg Whitton
Glyn Jones
Greg Lindstrom
Aidan Maccormick
Neil Bond
Ken Long
Lorelei Palmer-Willis
Kenneth Cox
Andrew Wilkin
Sonja Jordan
Alan O'Brien
Neil Guss

Ralph Armitage
Nick Monk
Nils Appenzeller
Dave Varo
Matt Clark
Itai Monnickendam
Nick Bramhall
Reini & Karl
Melanie Neethling
Tobias Richter
Stephen King
Renegade Scot
Craig Morris
Graham Duerden
Carmen Norman
Andrew Davis
Susan Betts
Anthony & Helen Fletcher
Robin & Jane Logie
Alban Fenle
Amy Lamond
Jan Horský
Michael Hirst
David Dear
Ryan Chubb
Gordon Forbes
Sylvia McGeer
Margaret & Antonio
Peter Cairns
Daniel Arnold
Jack Gamble
Iain Macleod
Ian Evans
Jane MacDougall

Margie Blankenship
Elgan Jones
Stefan Forster
Glen Sumner
Andy Bryant
Jayme Bell Armstrong
David McCrone
Peter Lindsay
Predrag Petrovic
Harsharn Gill
Frances Corbin
Guy Richardson
Matt Payne
Frank Williams
James Grant
Adrian Partridge
Michael Stirling-Aird
Kate Ings
Tim Parkin
John Henderson
Nikki & Pete
Jerome Oboils
Alistair Walker
Tim Nicholls
Ken Kelchtermans
Dylan Nardini
Craig Aitchison
Becky Bagnall
Bjørn Bäckstrøm Mæland
Graham Dunn
Ian & Marion Moncrieff
Phil Starkey
Jim Frost
Mark Littlejohn

Harvey Lloyd-Thomas
Scott Robertson
Josh Cooper
Pauline & Chris
Ian Brennan
Rafa Irusta
Glenn Wakeford
Andrew Whettam
Steve Deakin
Collette Patto
Ainsley, Louisa
Kristine, Sheena
Jackie Adair
The Togcast
Sam Gregory
Elaine Davies
Nigel Pullen
Andrew Greensted
Kalan Robb
Bryan Gray
Judith Parkyn
Michael Sugrue
Davor Perinović

Ross Lister Thorsten
Scheuermann
Linda Marshall
Jonathan &
Sarah Wilmshurst
Steve Forden
Mamraz Nagi
Stephan Fürnrohr
Mattias Sjolund
Tommy Dallmeyer
Roger Thomas
Esen Tunar
Sabin Simeonov
Simone Foedrowitz
Tim Middleton
Ian Turner
Ian Rhodes
Michael Stewart
Isla Elizabeth Mottram
Karl Seidl
Deirdre Murphy
Malcolm Robertson
Michael Sprock

Thank you to Fotospeed who provided the fine-art prints for the pre-orders and limited editions on their beautiful Platinum Etching paper.

Fotospeed
for every print

ABOVE: The Milky Way above Liathach

PREVIOUS: Camping on Suilven